Fundamentals of Biochemistry

Medical Course & Step 1 Review

a LANGE medical book

David DiTullio
Medical Student
David Geffen School of Medicine
University of California, Los Angeles
Los Angeles, California

Esteban C. Dell'Angelica, PhD
Professor
Department of Human Genetics
David Geffen School of Medicine
University of California, Los Angeles
Los Angeles, California

McGraw Hill Education

New York Chicago San Francisco Athens London Madrid Mexico City
Milan New Delhi Singapore Sydney Toronto

Fundamentals of Biochemistry: Medical Course & Step 1 Review

1 2 3 4 5 6 7 8 9 DSS 23 22 21 20 19 18

ISBN 978-1-259-64189-3
MHID 1-259-64189-9

Notice

Medicine is an ever-changing science. As new research and clinical experience broaden our knowledge, changes in treatment and drug therapy are required. The authors and the publisher of this work have checked with sources believed to be reliable in their efforts to provide information that is complete and generally in accord with the standards accepted at the time of publication. However, in view of the possibility of human error or changes in medical sciences, neither the authors nor the publisher nor any other party who has been involved in the preparation or publication of this work warrants that the information contained herein is in every respect accurate or complete, and they disclaim all responsibility for any errors or omissions or for the results obtained from use of the information contained in this work. Readers are encouraged to confirm the information contained herein with other sources. For example and in particular, readers are advised to check the product information sheet included in the package of each drug they plan to administer to be certain that the information contained in this work is accurate and that changes have not been made in the recommended dose or in the contraindications for administration. This recommendation is of particular importance in connection with new or infrequently used drugs.

This book was set in Minion Pro by MPS Limited.

The editors were Michael Weitz and Peter J. Boyle.

The production supervisor was Richard Ruzycka.

Production management was provided by Poonam Bisht, MPS Limited.

The designer was Mary McKeon.

The cover designer was Anthony Landi.

Library of Congress Cataloging-in-Publication Data

Names: DiTullio, David, editor. | Dell'Angelica, Esteban C., editor.

Title: Fundamentals of biochemistry : medical course & step 1 review /
    editors, David DiTullio, Department of Human, Genetics, David Geffen
    School of Medicine, University of California Los Angeles, Los Angeles,
    California, Esteban C. Dell'Angelica, PhD, Professor, Department of Human
    Genetics, David Geffen School of Medicine, University of California Los
    Angeles, Los Angeles, California.

Description: New York : McGraw-Hill, 2019. | Includes index.

Identifiers: LCCN 2018032155 | ISBN 9781259641893 (paperback) | ISBN
    1259641899 (paperback)

Subjects: LCSH: Biochemistry. | BISAC: MEDICAL / Dentistry / General.

Classification: LCC QD415 .F86 2019 | DDC 572—dc23

LC record available at https://lccn.loc.gov/2018032155

McGraw-Hill Education books are available at special quantity discounts to use as premiums and sales promotions or for use in corporate training programs. To contact a representative, please visit the Contact Us pages at www.mhprofessional.com.

Contents

Thank you for using *Fundamentals of Biochemistry: Medical Course & Step 1 Review*! Our goal is to help make biochemistry an approachable, clinically relevant subject for your first years of medical school and, most importantly, when you prepare to take the USMLE Step 1 examination.

Many medical students find biochemistry overwhelming. In most medical schools it is no longer a self-contained course like those taken in college—now it is one of many subjects covered in a whirlwind preclinical curriculum along with physiology, anatomy, pathology, and so on. In addition, at times it may seem hard to see the relevance of biochemistry for your future clinical career.

However, knowing biochemistry can greatly help you understand the basis of many disease processes. You do not have to memorize diseases and drugs by rote; you can attempt to group them by molecular mechanisms that make sense and reason through problems that show up months later, on Step 1 or on the wards. When you put biochemistry in a clinic context, learning it becomes less complicated as most puzzle pieces fall into place.

That is why we wanted to create a self-contained guide to high-yield biochemistry, with a focus on topics most likely to show up on the Step 1 exam. The text itself is in a simple outline format that contains all the high-yield information you need to know. Each chapter is also presented as a video lecture, available on McGraw-Hill's AccessMedicine site, so that you can review the topics in real time and add additional notes to the printed version.

You should be able to pick the book up at any chapter and find a self-contained summary of the relevant topic. We begin with the basics of the cell (Chapter 1) and of DNA and protein synthesis (Chapters 2–4), then cover the central aspects of metabolism (in Chapters 5–8), and finish with nutrition (Chapter 9) and genetics (Chapter 10). Where we felt that it may help you, we included cross-references among chapters. If you are not sure whether you need a full review of a topic, you can try the practice questions at the end of each chapter. Each has an explanation with reference to the text for quick review.

We hope that you will find this book useful as you go through your preclinical courses, so that you can put biochemistry in context of the rest of your studies. The book is designed to facilitate an efficient review of biochemistry as you begin to study for Step 1, as it covers all the high-yield topics that show up most often on that exam. We have found that, if you can begin your preparation for Step 1 having already studied some biochemistry, it will be much easier to review high-yield topics without getting overwhelmed.

Finally, we want to acknowledge some of the people who helped us put this resource together. Thanks a lot to the medical students at the David Geffen School of Medicine at UCLA, who generously gave their time to review and provide valuable feedback on each of the chapters. Thanks also to Michael and Peter at McGraw-Hill, for providing us with the chance and the freedom to create this book.

We hope that you will enjoy it!

Guide to Formatting

In this book, we highlight key terms using a few different formats, to help identify the most important facts to learn for the Step 1.

Blue bold text is used for key biochemistry terms, such as **microtubules**, **histones**, and so on. These are topics that are often tested on Step 1 and are worth paying extra attention to.

Red bold text is used for clinical correlates and disease, such as **Kartagener syndrome** or **aminoglycoside** antibiotics. These relate directly to the Biochemistry topics discussed in that section and often help connect Biochemistry to other topics across the Step 1 exam.

Regular bold text is also used to help emphasize high-yield facts, such as key symptoms of diseases, to help you quickly review sections or read through more efficiently.

We have tried to use this consistent structure so that whenever you use this book, whether the first time or for review later on, you can easily find what you're looking for and quickly identify high-yield information.

David DiTullio
Esteban C. Dell'Angelica, PhD

CHAPTER 1

The Cell

1. Purpose

    a. Biochemistry on the Step 1 focuses on cellular processes, so reviewing the basics of cell biology provides important foundation

    b. This is an introductory chapter, so focus on understanding the layout of the cell, especially how organelles are organized and their core functions

    c. Though some topics will come up in later chapters (as noted), focus on **clinical correlates:** Step 1 classically tests these basic concepts through Pharmacology

2. Overall cell structure

    a. Cells separate their key machinery into smaller membrane-bound **organelles**

    b. The most important components of the cell are

        1. **Plasma membrane** (regulates communication with the world outside of the cell)

        2. **Cytosol** (signal and material transport, protein synthesis)

        3. **Nucleus** (houses DNA, transcription)

        4. **Endoplasmic reticulum** and **Golgi** (protein processing and transport)

        5. **Mitochondria** (site of energy production, other metabolic processes)

        6. **Lysosome** (breakdown of waste)

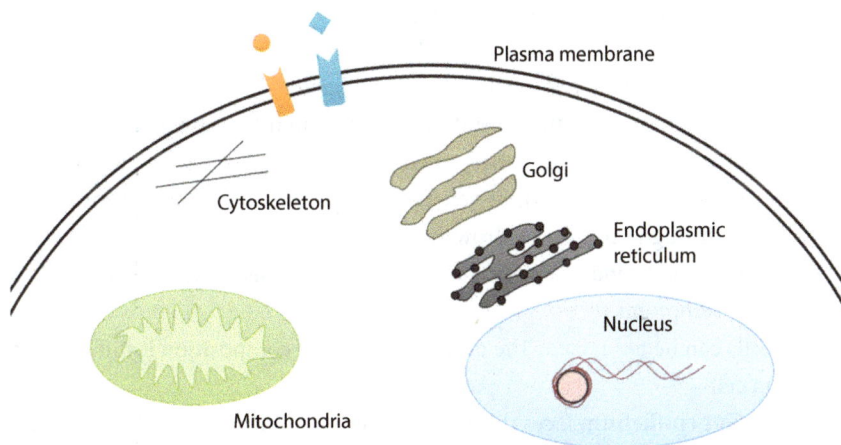

Figure 1-1. Overview of cell structure.

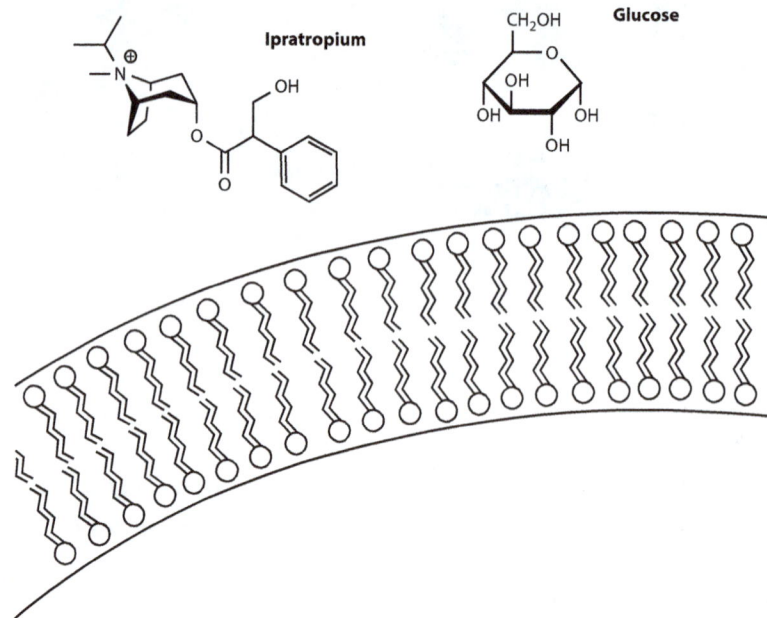

Figure 1-2. Plasma membrane.

3. Plasma membrane

a. The plasma membrane is the point of contact with the outside of the cell

b. It is made of a phospholipid bilayer that prevents cell contents from spilling out and regulates cellular intake—so it is important for Pharmacology

c. Hydrophobic molecules can pass through the membrane freely

   1. Steroid hormones and thyroid hormone are common hydrophobic compounds that pass through the plasma membrane

d. Hydrophilic and large molecules require protein channels or transporters to enter the cell

   1. **Glucose transporters** (GLUTs) respond to insulin differently depending on isoform
      - GLUT4 is insulin responsive (muscle, adipose tissue)
      - Some diabetes drugs target glucose/sodium cotransporters (SGLTs)
      - These will be reviewed in detail in Chapter 5

   2. Drugs only enter the central nervous system if they are hydrophobic due to the blood-brain barrier (BBB), composed of tightly packed cells
      - Antihistamines—first-generation drugs are nonpolar and pass through the BBB → drowsiness
      - Second generation are charged and remain peripheral → no CNS symptoms

e. Cells can be polarized: The plasma membrane composition is different on cell

   1. **Gut epithelium** faces the gut lumen on one side and blood vessels on the other, so each side has unique signals and transporters
      - Lactase is only expressed on the luminal side

   2. **Embryologic development** also relies on polarity of cells to help develop and shape the organism

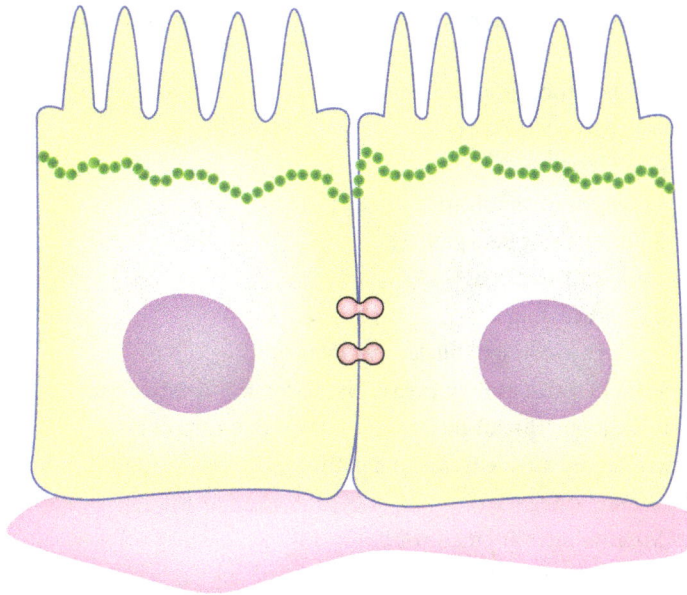

Figure 1-3. Membrane polarity.

4. Cytosol

   a. The cytosol is a busy place, so it is important to focus on what specifically
       is relevant for Step 1
       1. How signals from outside the cell are processed in the cytosol vs.
           nucleus (Chapter 4)
       2. How protein synthesis is localized between cytosol and endoplasmic
           reticulum (Chapter 3)
       3. What parts of metabolism happen in the cytosol (Chapter 5)
       4. The structure and function of the cytoskeleton, the scaffolding that
           gives cells their shape and allows for intracellular transport
   b. **Signal transduction:** Signals from outside the cell are relayed to their
       destinations in different ways
       1. Initiate cytosolic signaling cascades
       2. Diffuse directly to the nucleus

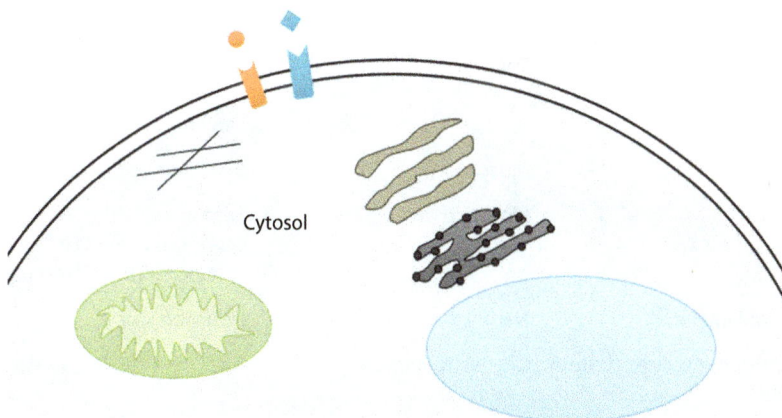

Cytosol

Figure 1-4. Cytosol.

        3. Diffuse to the cytosol and bind receptors → translocate to the nucleus together

  c. Metabolic processes: Step 1 sometimes asks about where key reactions take place. Many pathways take place in the cytosol
        1. Glycolysis
        2. Glycogen metabolism
        3. Pentose phosphate pathway
        4. Fatty acid synthesis

  d. mRNA processing
        1. RNA processing in nucleus vs. cytosol
        2. Protein translation in cytosol vs. endoplasmic reticulum

  e. Cytoskeleton: Thread-like proteins throughout the cytosol provide structure and enable trafficking pathways within the cell; you should know the 3 main categories
        1. Microtubules are made of tubulin
          • Highways to move things through the cell
          • Cilia and flagella: Sensing extracellular environment, cell motility
        2. Microfilaments (actin filaments) are made of actin
          • Actin filaments act in a central role in sarcomeres, the unit of muscle cell contraction
        3. Intermediate filaments provide structure and vary by cell type
          • Their diversity by cell type makes them useful in characterizing cancer subtype

  f. Cytoskeletal proteins are tested on Step 1 via 3 major categories
        1. Microtubule dysfunction leads to Kartagener syndrome
          • *Clinical:* Infertility, respiratory problems, situs inversus
        2. Drugs target microtubules to prevent cell motility and cell division
          • Antimicrobial drugs (e.g., mebendazole)
          • Anticancer drugs (e.g., vincristine)
        3. In cancer, intermediate filament stains identify cell of origin
        4. These will be discussed in more detail in Chapter 4

Table 1-1. Intermediate filament stains.

Stain	Cell type	Used to identify
Cytokeratin	Epithelium	Epithelial tissue, e.g. skin, GI, ovarian, squamous cell carcinoma
Desmin	Muscle	Rhabdomyosarcoma, etc.
Vimentin	Mesenchymal tissue	Sarcomas, endometrial carcinoma, renal cell carcinoma, meningioma
Neurofilament	Neurons	Neuroblastoma
Glial fibrillary acidic protein (GFAP)	Glia (astrocytes)	Astrocytoma, glioblastoma

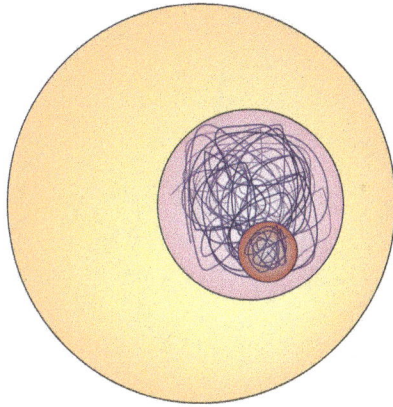

Figure 1-5. Nucleus.

5. Nucleus

    a. The nucleus houses genomic DNA, separated into 23 pairs of chromosomes

    b. It is bound by a porous, double lipid membrane, which regulates how signals get in and out

    c. DNA is stored in the nucleus as chromatin, tightly wrapped around proteins called histones

    d. Processes that happen here include

        1. Transcription factor interaction with and binding to DNA

        2. DNA synthesis

        3. RNA synthesis

6. Endoplasmic reticulum (ER)

    a. The ER is divided into rough and smooth

        1. **Rough ER** is the classic site of protein synthesis

        2. **Smooth ER** is the site of steroid synthesis and drug/toxin breakdown

    b. From the ER, proteins are trafficked to the Golgi apparatus and eventually to their final destination

    c. The rough ER is especially prominent in **plasma cells** of the immune system, which synthesize large quantities of secreted antibodies

    d. The smooth ER is especially prominent in **liver cells**, which play a key role in detoxification and steroid synthesis

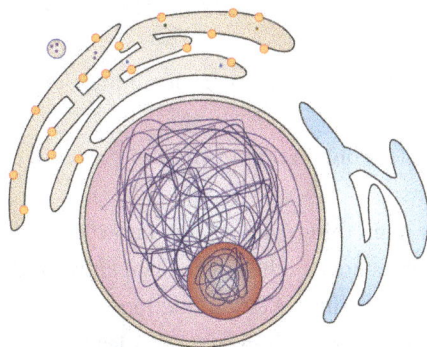

Figure 1-6. Endoplasmic reticulum.

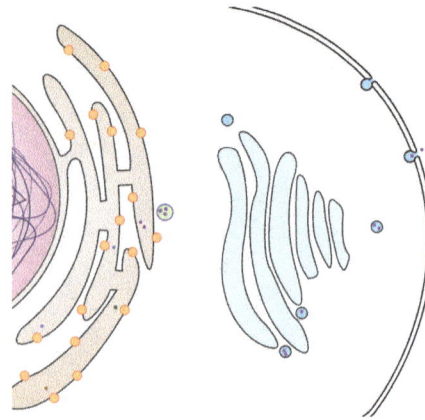

Figure 1-7. Golgi apparatus.

7. Golgi

a. In the rough ER and Golgi, post-translational modifications of newly made proteins take place

b. One key feature of post-translational modifications is tagging proteins with markers that direct them to the correct destination

   1. Errors in this process can lead to severe disease due to mistargeting of a functional protein to the wrong compartment

c. The Golgi apparatus is a series of membrane compartments that processes proteins after synthesis and directs them to their final destination

   1. *Cis*-Golgi (closest to ER) to *trans*-Golgi (farthest from ER)

d. Proteins are transported through the Golgi compartments, and to other organelles, in specially coated vesicles

e. The main thing to remember about the Golgi is that often proteins have to be processed in the Golgi to function correctly; 2 main aspects are often tested

   1. Mannose-6-phosphate is key "tag" that identifies proteins to be trafficked to the lysosome specifically (example of post-translational modification)

   2. Collagen synthesis requires multiple processing steps of the primary protein to form into its functional structure (Chapter 3)

8. Mitochondria

a. Mitochondria are the energy producers of the cell—they convert nutrients into energy

b. Mitochondria have inner and outer membranes, creating 2 distinct compartments

c. Mitochondria also contain their own mini-genomes that encode for some of their proteins

d. For Step 1, the main things you need to know are

   1. Which key metabolic pathways are happening in the mitochondria

   2. How poisons can affect ATP production

   3. How mutations in mitochondrial DNA lead to severe genetic disease

   4. We will cover these in detail in Chapters 5 and 10

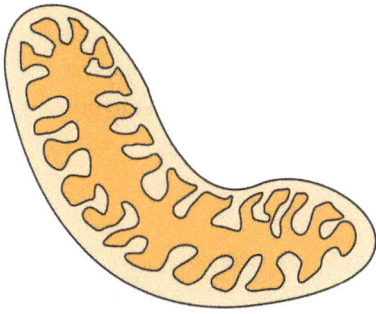

Figure 1-8. Mitochondria.

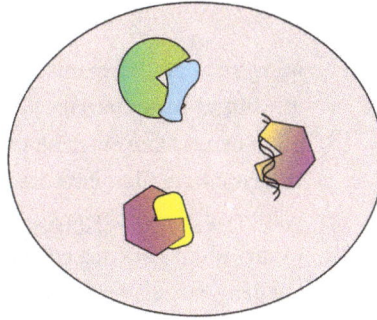

Figure 1-9. Lysosome.

9. Lysosome

   a. Lysosomes are the "trash and recycling" centers of the cell
   b. Lysosomes are highly acidic, membrane-bound compartments that degrade many types of compounds
   c. This is accomplished through special lysosomal proteins, which are tagged in the Golgi with mannose-6-phosphate to direct them to the lysosome
       1. In I-cell disease, failure to tag proteins correctly leads to dysfunctional lysosomes and buildup of these proteins in the bloodstream
       2. There are also mutations within specific lysosomal enzymes that can result in failed degradation of junk products in the lysosome, leading to lysosomal storage diseases (see Chapter 10)

10. Peroxisome

   a. The peroxisome is similar to the lysosome in many ways, although it does not come up in the Step 1 as often
   b. Specifically, primarily very long chain fatty acids and other complex fatty acids are broken down by oxidation (see Chapter 8)
   c. Disease relevance: Like the lysosome, deficiencies in specific peroxisome enzymes lead to problems in breaking down these compounds

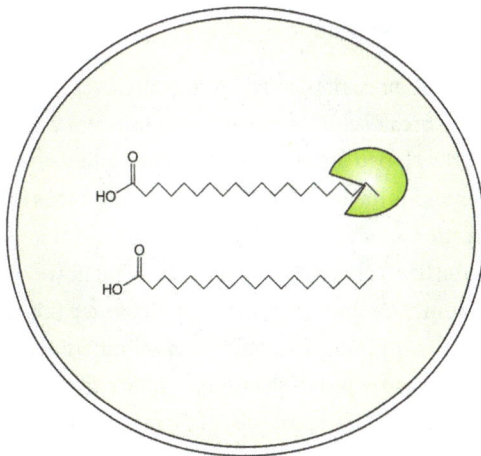

Figure 1-10. Peroxisome.

11. Disease relevance summary

a. Many of the diseases introduced here will be covered in later chapters

b. The **blood-brain barrier** utilizes plasma membrane properties to prevent polar drugs from entering the CNS

c. **Intermediate filaments** are used to test for specific cell types, as in identifying cancer type

d. **Microtubules** are defective in diseases such as Kartagener syndrome, and are **drug targets** for both antimicrobial and anticancer treatments

12. Wrap-up

a. You should have a solid understanding of the cell makeup before studying biochemistry in more depth, since every process relates to the geography of the cell

b. This chapter is a very brief introduction to the topics that will come up throughout the rest of the text and put cell processes in context

c. Review the cytoskeleton and plasma membrane here, but diseases related to each organelle will be reviewed in more detail in the following chapters

Practice questions

1. A pediatric endocrinologist conducts a new patient exam on a 3-year-old boy who exhibits unusual symptoms. His parents state that they carefully monitor his diet and give him a daily multivitamin as directed, but despite their care, he has continued to gain weight and is now at the 99th percentile of the weight curve. On exam, he has puffy facial features and appears lethargic. Furthermore, basic cognitive tests indicate he is developmentally delayed. Based on these symptoms, the physician worries about a thyroid hormone deficiency. Which of the following proteins is most critical for thyroid hormone signaling?

A. Blood-brain barrier (BBB) integrity protein
B. Mitochondrial receptor
C. Nuclear receptor
D. Plasma membrane channel protein
E. Plasma membrane pump
F. Post-translational modification protein

2. A 16-year-old patient presents to her primary care physician after returning home after a spring break school trip to South America 2 weeks ago. Soon after she returned, she developed a runny nose and persistent sneezing. She has been getting headaches and is having trouble sleeping. She states she is worried about a respiratory infection and says that school friends do not want to be near her because they say she has "tuberculosis." The physician, however, suspects seasonal allergies. He recommends she take an over-the-counter (OTC) antihistamine to see if it improves her symptoms. He recommends loratadine, a second-generation antihistamine, for daytime, and diphenhydramine, a first-generation drug of the same class, for nighttime. Which of the following statements is most likely true about these drugs given his recommendations?

A. Diphenhydramine blocks neuronal signaling more readily than loratadine
B. Diphenhydramine is metabolized by the liver more rapidly than loratadine

C. Loratadine crosses the blood-brain barrier more readily than diphenhydramine

D. Loratadine has a higher effective dose than diphenhydramine

E. Loratadine is more lipophilic than diphenhydramine

3. A 28-year-old graduate student presents to the university's student health clinic. She complains of abdominal pain and persistent flatulence. She has not had these symptoms before. When asked about her diet, she states that she used to be vegan, but recently elected to become vegetarian as she had difficulty maintaining a balanced diet because of her busy schedule. The physician suspects lactose intolerance and recommends alterations to the patient's diet that might help. When the patient requests that a biopsy be taken, the physician explains why this is not yet necessary and what the biopsy would likely show. Where would the affected area appear, using an imaging technique appropriate for the condition?

A. Epithelial cell cytosol

B. Epithelial cell mitochondria

C. Endothelial cell basal membrane

D. Hepatocyte cytosol

E. Luminal epithelial membrane

4. A 58-year-old woman is referred to a neurologist by her primary care provider for evaluation of headaches and dizziness that have worsened over the past month. She states that she had migraines in her 20s, for which she took topiramate prophylactically and sumatriptan as needed. However, she has not had episodes in the past several years and stopped taking her medication after menopause with no adverse symptoms. After conducting her exam, the neurologist orders a contrast-enhanced MRI. Imaging shows an irregular, hyperintense region in the right temporal lobe. The doctor takes a biopsy of the tissue in that area, and the pathology report notes staining for a particular substance that allows the neurologist identify astrocytes as the cell of origin. Which of the following stains was most likely positively identified in the pathology specimen?

A. Cytokeratin

B. Desmin

C. Glial fibrillary acidic protein

D. Neurofilaments, light subunit

E. Peripherin

5. A 33-year-old man presents to the urgent care clinic complaining of intermittent abdominal pain and diarrhea for the last month. His only recent travel was a backpacking trip through the Amazon rainforest, from which he returned 2 months ago. He states his first bout of GI symptoms began a few weeks after returning from his trip, which he attributed to "readjusting to his normal lifestyle." Suspecting a microbial or parasite infection, the physician collects a stool sample, and provides fluid support to the patient to treat intervening dehydration while waiting for the results. The stool sample shows presence of larvae, and the physician calls the patient to inform him that he has been prescribed mebendazole. This drug therapy directly affects which of the following molecules?

A. Actin

B. ATP synthase

C. DNA

D. Myosin

E. Ribosomes

F. Tubulin

6. A 42-year-old patient is brought to the emergency room by ambulance on suspicion of attempted drug overdose. The patient's girlfriend found the patient unresponsive in their apartment. Upon questioning, she discloses that he had been regularly taking various opioids since a back surgery 3 years ago, and had continued after his prescription ran out by obtaining the medication online without a prescription. She has the pill bottle she found, which the physician is able to identify as Vicodin, an opioid/acetaminophen formula. The physician administers naloxone and respiratory support, and the patient is stabilized. However, a blood panel demonstrates AST and ALT both over 1,000 IU/L. From what part of the hepatocytes does the toxicity arise?

A. Golgi apparatus

B. Lysosome

C. Mitochondria

D. Nucleus

E. Peroxisome

F. Endoplasmic reticulum

DNA Structure & Synthesis

CHAPTER 2

1. Purpose

a. This chapter covers DNA structure, synthesis, and repair

b. There is a lot of information here; consider how much time you can devote to studying the material and prioritize information based on how much you can reasonably cover

c. Some topics are particularly high yield; usually it is because there are diseases or drugs that involve these pathways

   1. Enzymes involved in DNA synthesis
   2. Types of DNA mutations
   3. DNA repair mechanisms and associated diseases
   4. Nucleotide synthesis and diseases

2. Nucleotide bases

a. DNA and RNA, which carry the code for making proteins, are made up of individual bases

   1. Adenine (A)
   2. Guanine (G)
   3. Cytosine (C)
   4. Thymine (T)—DNA only
   5. Uracil (U)—RNA only

b. Purine bases (A, G) have 2 rings while pyrimidine bases (C, T, U) have just 1

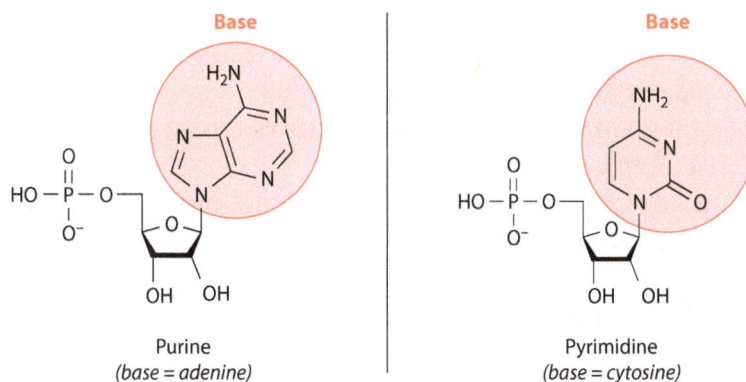

Figure 2-1. Basic nucleotide structure.

Figure 2-2. DNA and histones.

   c. DNA base pairing depends on nucleotide properties
     1. To keep DNA size uniform must have 1 purine + 1 pyrimidine
     2. A and T/U pair together with 2 hydrogen bonds
     3. C and G pair together with 3 hydrogen bonds
     4. This means that C-G rich regions of DNA are highly stable (CpG islands)

3. DNA structure

   a. DNA strands are composed of bases along a phosphate-sugar backbone
     1. The backbones run in opposite directions, and DNA is read along the backbone ($5' \rightarrow 3'$)
     2. DNA is negatively charged
   b. During storage, DNA wraps around histone proteins — positively charged proteins (lysine, arginine)
   c. This histone-wrapped DNA is stored in a highly condensed structure called chromatin
     1. Heterochromatin is tightly packed and generally inactive
     2. Euchromatin is unwound and generally active

4. Epigenetics

   a. The DNA sequence determines a protein's sequence (primary structure), but is just one piece of biological variation: Regulation of expression is also important
   b. Epigenetics describes this regulation that goes beyond genetics (DNA sequence)
   c. For Step 1, the main players in epigenetics are methylation and acetylation of DNA and histones
   d. DNA methylation occurs in 2 situations
     1. General methylation of DNA helps the cell distinguish template vs. new strands during replication
     2. CpG islands are methylated to repress expression
   e. Histone modification affects how accessible genes are → alters transcription rates
     1. Methylation usually mutes transcription
     2. Acetylation activates transcription (removes + charge from histones)

f. Methylation of chromosomes underlies imprinting, where paternal or maternal genes are turned off
   1. Patients are heterozygous but the normal copy is silenced
   2. Prader-Willi syndrome (maternal silencing, mutation inherited from father)—excessive eating
   3. Angelman syndrome (paternal silencing, mutation inherited from mother)—excessive laughter, smiling

5. DNA replication

a. The replication process is important because Step 1 likes to test ways in which errors can lead to DNA mutations, which underlie genetic disease and cancer
b. Less important to describe the process from start to finish than to identify the function of specific proteins
   1. **Helicase:** Unwinds DNA strand to allow synthesis on each strand individually
   2. **Topoisomerase:** Similar to helicase, helps to "unpack" and "repack" DNA into tight coils by nicking to release extra coils
      - Prokaryotic inhibitors (DNA gyrase): Fluoroquinolones (e.g., ciprofloxacin)
      - Eukaryotic inhibitors: Etoposide, teniposide
   3. **RNA primase:** Adds short RNA strands to serve as template for synthesis of new strand
   4. **DNA polymerase:** Synthesizes new DNA strand from 5′ to 3′
   5. **Ligase:** Joins DNA fragments together after synthesis from multiple origins

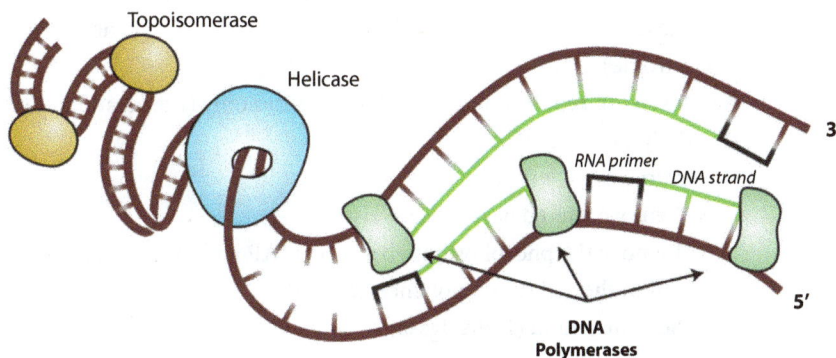

Figure 2-3. DNA replication.

c. Some key aspects of the process are tested
   1. Synthesis always happens 5′ to 3′ which means synthesizing "backwards" from the origin is impossible: Multiple **Okazaki** fragments are synthesized 5′ to 3′
   2. Prokaryotes have special types of DNA polymerase
      - DNA Pol III is standard, can synthesize new bases 5′ to 3′ but "erase" in 3′ to 5′ to undo errors
      - DNA Pol I can erase forward and backward, so it can erase the RNA primer and replace with DNA
   3. Parent DNA strands are methylated at some C and A bases; helps identify mutations

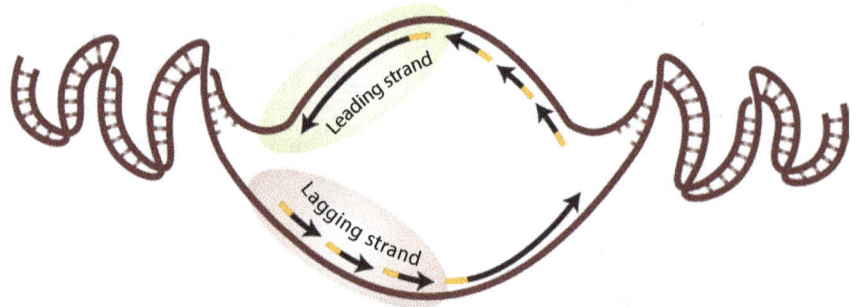

Figure 2-4. Leading and lagging strands.

6. DNA damage and repair

a. Alterations to the DNA sequence can happen during synthesis or from environmental damage, or result from enzyme deficiency

   1. If not repaired, accumulated DNA damage can increase the risk of cancer

b. Types of DNA damage relates to how mutations affect codon translation (Chapter 3)

c. Know 5 key pathways for Step 1, at different severity levels of damage

   1. **Mismatch repair:** Incorrect pairing of bases during synthesis
      - Repair: New strand hasn't been methylated yet so mismatched nucleotide can be removed and replaced with the correct base
      - Diseases: **Hereditary nonpolyposis colorectal cancer** (Lynch syndrome)

   2. **Base excision repair:** Small mutations of bases only (e.g., deamination) that do not distort DNA
      - Repair:
         - Remove altered base (glycosylase),
         - Remove the phosphoribose backbone (AP endonuclease, lyase),
         - Fill in the gap (DNA polymerase), and
         - Seal the strand (DNA ligase)
      - No specific diseases associated

   3. **Nucleotide excision repair:** Larger mutations distort DNA structure (external damage, e.g., UV)
      - Repair:
         - Remove damaged section (endonucleases),
         - Fill the gap (DNA polymerase), and
         - Seal the strand (DNA ligase)
      - Diseases: **Xeroderma pigmentosum** (pyrimidine dimers)

   4. **Homologous recombination:** Double-strand breaks create entire breaks in DNA strand; cell needs to reattach
      - Repair: Homologous chromosome is used as a template to identify site of break and fill in any lost DNA
      - Diseases: *BRCA1* and *BRCA2* encode genes involved in this pathway; mutations increase lifetime risk of breast cancer

Table 2-1. Types of DNA mutations.

Mutation	Effect	Details and examples
Nucleotide changes		
Silent	No change to protein sequence	Many single nucleotide polymorphisms (SNPs)
Missense	Change one amino acid to another	*Conservative:* aa has similar properties *Nonconservative:* aa with different properties; e.g., sickle cell (Glu → Val)
Nonsense	Premature stop codon	Truncated protein Duchenne muscular dystrophy (some forms)
Amino acid changes		
Frameshift	Insert/Remove bases (non-multiple of 3)	Changes downstream amino acid sequence → effect similar to nonsense Duchenne muscular dystrophy (some forms) Tay-Sachs disease (some forms)
Insertion/Deletion	Insert/Remove amino acids	Variable severity depending on location Cystic fibrosis (deletion of F508 → misfolding)

5. **Nonhomologous end joining:** Double-strand breaks need to be repaired ASAP to minimize loss of bases at the break ends
   - Repair: When homologous recombination not available, just paste the 2 ends back together; very error prone but better than nothing
   - Diseases: **Ataxia telangiectasia** (multisystem) and **Fanconi anemia**

7. Nucleotide synthesis and salvage

a. Step 1 will test pieces of nucleotide synthesis and salvage pathways—focus on the steps that have disease mutations or drugs targeting them

b. Nucleotides can be created by *de novo* synthesis or salvaged from degraded DNA

1. *De novo* nucleotide synthesis provides targets of many **cancer and autoimmunity drugs** because these cells have to grow very quickly

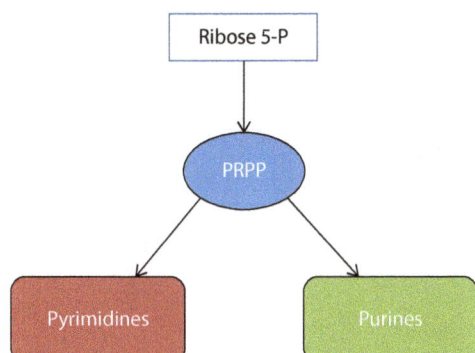

Figure 2-5. Core *de novo* nucleotide synthesis pathway.

2. Salvage is central to gout pathology and **inherited diseases** because DNA breakdown leads to toxic metabolites

8. *De novo* synthesis

   a. Phosphoribosyl pyrophosphate (PRPP) is a central building block in *de novo* nucleotide synthesis
   b. Purine synthesis: Synthesized to IMP first and then to GMP and AMP
   c. Pyrimidines: Orotic acid (base) is synthesized first and then added to PRPP
      1. Orotic acid, an intermediate base, is combined with PRPP to form uracil monophosphate (UMP)
      2. UMP is then converted into thymine- and cytosine-containing nucleotides
      3. Synthesis of thymine nucleotides starts with conversion of the uracil nucleotide UDP to its deoxynucleotide form (dUDP) via ribonucleotide reductase
      4. Conversion of uracil to thymine requires the dihydrofolate reductase system to add a methyl group
         • This step requires **folic acid**—important in pregnancy, blood cells
   d. Diseases of *de novo* synthesis: Orotic aciduria
      1. Enzyme deficiency results in **orotic acid** buildup, presence in the urine
      2. **Mental retardation** and **stunted growth** are present in severe cases
      3. Treat when necessary with downstream product: Uridine
   e. Drugs targeting *de novo* synthesis inhibit cell growth (vs. cancer, autoimmunity, and infection)
      1. Drugs are either base analogs or specific enzyme inhibitors
      2. **Purine drugs:** 6-mercaptopurine; mycophenolate, ribavirin
      3. **Pyrimidine drugs:** Leflunomide, 5-fluorouracil, dihydrofolate reductase inhibitors (methotrexate, trimethoprim, pyrimethamine)
      4. **Both:** Hydroxyurea

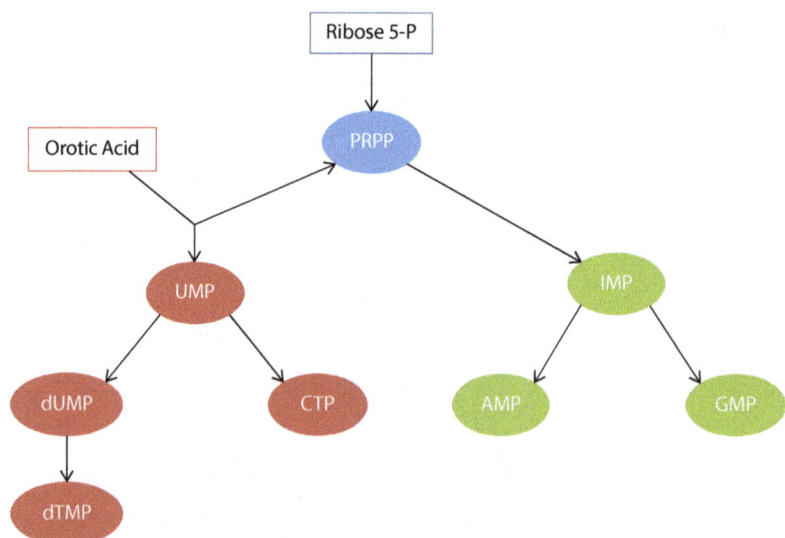

Figure 2-6. Key pieces of the *de novo* synthesis pathway.

Table 2-2. Drugs targeting nucleotide synthesis.

	Purine	Pyrimidine	Nonspecific
Nucleotide analogs	Azathioprine 6-mercaptopurine Ribavirin	5-fluorouracil	
Other enzyme inhibitors	Mycophenolate	Leflunomide Methotrexate Trimethoprim	Hydroxyurea

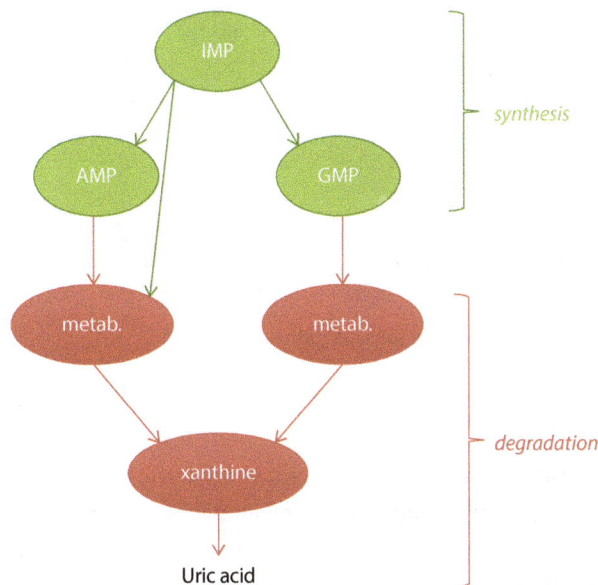

Figure 2-7. Basic purine metabolism pathway.

f. Regulation of nucleotide synthesis
   1. Nucleotides levels are regulated by coordinated feedback mechanisms
   2. All DNA bases inhibit formation of their precursors (IMP/orotic acid)
   3. ATP activates orotic acid synthesis (make more pyrimidines, fewer purines)

9. Purine catabolism and salvage

a. Defective degradation of purine nucleotides underlies important disease pathologies
b. Xanthine is the central metabolite and is converted into uric acid for renal excretion
c. AMP and IMP share a general degradation pathway, while GMP follows a separate pathway on the way to xanthine
d. The purine salvage pathway can be thought of in many ways as the reverse of its corresponding *de novo* synthesis pathway
e. There are 3 diseases related to purine degradation and salvage: Two are due to enzyme deficiencies, one is due to excessive turnover

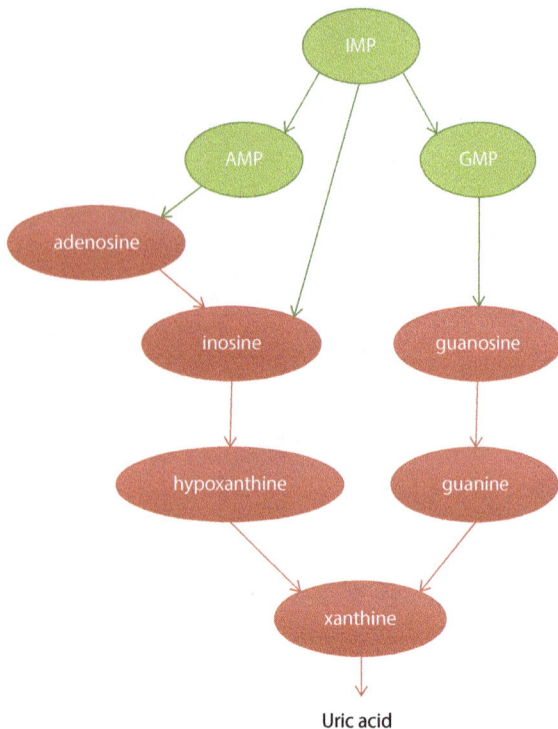

Figure 2-8. Purine metabolism pathway in detail.

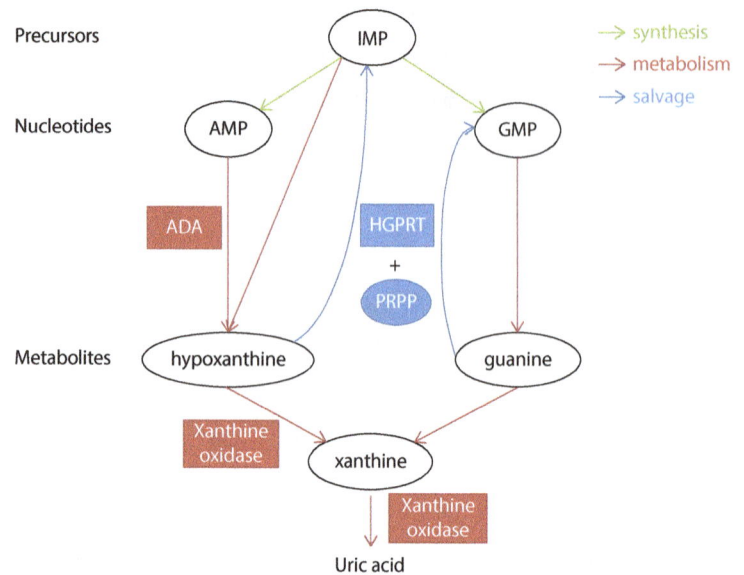

Figure 2-9. Purine salvage pathway in detail.

1. Excessive purine breakdown leads to high uric acid, which accumulates in joints as crystals and causes inflammation → **gout**
   - Treat by preventing breakdown: Allopurinol and febuxostat inhibit xanthine oxidase
   - Probenecid works on the opposite side, promoting removal of uric acid that has been formed
2. **HGPRT** deficiency leads to **Lesch-Nyhan syndrome**
   - X-linked recessive disease, primarily affecting males
   - Symptoms include mental retardation, self-mutilation and aggression, and gout
   - Treat by preventing uric acid formation
3. **Adenosine deaminase** deficiency leads to **severe combined immunodeficiency (SCID)**

10. Disease relevance summary

a. **Methylation** of DNA and histones silences gene expression, underlying Angelman and Prader-Willi syndromes

b. **DNA synthesis inhibitors** are used to treat infections and cancer by inhibiting cell proliferation

c. **DNA mutation** severity roughly correlates to disease severity
   1. Missense mutations and single-aa deletions lead to impaired function
   2. Frameshift and nonsense mutations often lead to loss of protein function

d. Mutations in **DNA repair pathways** lead to cancer
   1. Double-strand breaks are more damaging than single-strand mutations

    e. *De novo* **nucleotide synthesis** is a crucial cell function

        1. Anti-cancer and inflammation drugs target this process, including 6-mercaptopurine, hydroxyurea, methotrexate, and trimethoprim being most common

        2. Orotic aciduria is a deficiency of pyrimidine synthesis resulting in orotic acid buildup

          • The hallmark symptom of orotic aciduria is presence of orotic acid in the urine

          • Treatment with uridine bypasses the defective step

    f. **Purine catabolism and salvage** deficiencies include **SCID** and **Lesch-Nyhan,** respectively, and anti-gout drugs allopurinol, febuxostat, and probenecid prevent toxic metabolite buildup

11. Wrap-up

    a. Prioritizing information is key in this chapter

    b. Begin with the pharmacology, as drugs target DNA manipulation in many contexts

    c. Next, know commonly tested diseases (e.g., Lesch-Nyhan)

    d. Finally, be able to generally describe specific mechanisms (e.g., DNA replication, mismatch repair)

Practice questions

1. A pathologist is examining a patient sample sent from a medicine clinic. The pathologist examines the cellular morphology, and as part of the education of new pathology residents, also images several of the cells in various locations using electron microscopy and a special stain against chromatin. In one set of cells, she notes lighter staining of chromatin and describes to the residents that this represents euchromatin. What does this observation imply about that region of chromatin?

    A. It is highly methylated and transcriptionally active

    B. It is highly methylated and transcriptionally inactive

    C. It is relatively unmethylated and transcriptionally active

    D. It is relatively unmethylated and transcriptionally inactive

    E. It was stained while DNA was being replicated

2. An 8-year-old boy is brought to the pediatrician's office by his parents for a normal checkup. The boy was previously diagnosed with an unspecified intellectual disability without a determined cause. He is 120 cm (47 in.) tall and weighs 50 kg (110 pounds); his BMI is 34.7 kg/m². Realizing that his obesity may be a symptom of a larger genetic disease, the physician tests for Prader-Willi syndrome. The test finds that the patient is heterozygous for a deletion in 15q11-q13, sharing the same deletion variant with his father, who is unaffected. What is the most likely explanation for the difference in phenotype given the same genetic background for these 2 individuals?

    A. Methylation of the maternal chromosome

    B. Methylation of the paternal chromosome

    C. Nondisjunction event in maternal meiosis

    D. Nondisjunction event in paternal meiosis

    E. Spontaneous mutation in the embryonic stage

3. A 28-year-old woman comes into the genetics clinic with her long-term boy-friend to inquire about genetic testing options. She and her partner hope to have a child, but she is concerned because her boyfriend's family has a history of Huntington's disease. He does not know if he has the disease mutation and is not interested in finding out. However, the couple wants to ensure that their child will not be affected. The geneticist agrees to help the couple with the use of pre-implantation diagnosis. In this procedure, DNA will be isolated from single cells in the early morula stage and analyzed by PCR. Activity of which of the following proteins is important to ensure access of PCR enzymes to DNA?

   A. DNA polymerase
   B. Helicase
   C. Histone H1
   D. RNA polymerase
   E. RNA primase

4. A pulmonologist studying cystic fibrosis begins a study to examine the link between genetic mutation and disease severity. DNA samples have been col-lected from all cystic fibrosis patients that have come to the clinic over the past 5 years. The researcher sequences the *CFTR* gene from each of these patients and correlates it to their lab values, clinical symptoms, and other outcome measures of disease progression. Which of the following mutations would be most likely to lead to translation of a protein shorter than non-mutated CFTR protein?

   A. Frameshift mutation in intron 1
   B. Insertion of "ATG" after codon 178 of exon 5
   C. Missense mutation at codon 560 of exon 11
   D. Nonsense mutation at codon 39 of exon 2
   E. Silent mutation at codon 5 of exon 1

5. A 12-year-old girl is taken to the beach by her parents while on vacation. After going swimming in the ocean, she does not put on more sunscreen, and she comes home with a sunburn on a significant portion of her skin. What DNA process is most likely taking place to address the DNA damage in her cells?

   A. Base-excision repair
   B. Homologous recombination
   C. Mismatch repair
   D. Nonhomologous end joining
   E. Nucleotide excision repair

6. Cancer biologists working for a pharmaceutical company are conducting research on new antimetabolites that could prove effective at treating rapidly dividing cancer cells. They are focusing on nucleotide synthesis, as the rate of base synthesis is closely correlated to rate of replication. A drug designed as an inhibitory analog of which of the following compounds would be expected to most effectively inhibit nucleotide production?

   A. Inosine monophosphate
   B. Hypoxanthine
   C. Orotic acid
   D. Phosphoribosyl pyrophosphate
   E. Uridine monophosphate

7. A 2-week-old newborn boy is brought to the emergency room on Christmas Day by his parents. They explain that he developed a significant cough on Christmas Eve, and when they called the urgent care hotline, they were told to wait a day to see if it worsened. On Christmas Day, their son had a severe cough, lethargy, and was refusing to eat. The pediatric emergency medicine physician quickly realizes that there is a serious problem. He consults with the on-call pediatrician, and together they make a diagnosis of severe combined immunodeficiency (SCID), a life-threatening immunodeficiency. Which of the following represents the most likely pathologic mechanism of this disease?

   A. Accumulation of adenosine nucleotides
   B. Accumulation of uric acid
   C. Failure of the *de novo* purine synthesis pathway
   D. Inability to convert UMP into downstream products
   E. Overproduction of guanine nucleotides

8. An 81-year-old man complains to his primary care physician about pain in his joints. His large toes in particular, are especially painful. He concedes that he drinks about 2 beers every night, and still eats red meat despite his physician recommending he cut back. The physician explains that he is probably suffering from gout, which is caused by infiltration of the joints by crystals formed during purine metabolism. If the physician is to prescribe a drug to treat this condition, which of the following enzymes should the drug affect?

   A. Adenosine deaminase
   B. Hypoxanthine-guanine phosphoribosyltransferase
   C. Phosphoribosyl pyrophosphate synthetase
   D. Ribonucleotide reductase
   E. Xanthine oxidase

9. A 3-year-old boy and his 15-month-old brother are brought to the pediatric urgent care clinic by their mother. This is their first visit at this clinic. She states that, while she took her older son to regular doctor's visits up to 6 months, she did not have health insurance until recently so did not bring him back since then. However, she is concerned now, since both children have begun acting out. She's noticed scratches and bite marks on both of them and worries they're fighting, and also seems to have some sort of "twitching disorder." Even worse, the younger son seems to be developing the same tic. She is worried about Tourette syndrome, and asks if there is a genetic test available. The pediatrician conducts a physical exam, and orders urine samples from each of the patients. He suspects a genetic disorder impeding DNA metabolism. If the physician is correct, which of the following is the best treatment option?

   A. Adenosine
   B. Febuxostat
   C. Hydroxyurea
   D. 6-Mercaptopurine
   E. Methotrextate

10. A 12-year-old boy has been diagnosed with acute lymphoblastic leukemia (ALL). His oncologist is considering chemotherapy regimens. After looking at recommended treatment protocols, the oncologist identifies an appropriate

course of treatment, including vincristine and prednisone. At the end of the treatment, the oncologist recommends continuing maintenance therapy with a less aggressive chemotherapeutic. She explains a drug that can be taken orally, which preferentially inhibits growth in quickly dividing cells by targeting DNA synthesis. In particular, it inhibits proteins that help synthesize the nucleotides required for DNA. Which of the following drugs is she most likely describing?

A. Ciprofloxacin
B. Colchicine
C. Folate
D. 6-Mercaptopurine
E. Probenecid

Transcription & Translation

CHAPTER

3

1. Overview

  a. Similar to DNA structure and synthesis, these subjects are closely related to Hematology/Oncology and Microbiology sections
  b. Thus, much of the focus is, as usual, on pharm and key disease applications (e.g., sickle cell anemia)
  c. In addition, 2 Biochemistry-specific topics that tend to show up on Step 1: Collagen synthesis and the *lac* operon

2. Regulation of gene expression

  a. There are several ways gene expression is regulated
    1. Initiation requires transcription factor binding to **promoters** (directly upstream of the gene)
       - Promoters are often cell-type specific
       - TATA and CAAT boxes are common features of promoters
    2. Expression is regulated by **enhancers** and **silencers** (anywhere around the gene)
    3. Methylation of DNA (CpG islands) mutes expression (see Chapter 2)
    4. Histone modification also regulates expression
       - Methylation of histones usually mutes expression
       - Acetylation of histones activates expression
  b. The *lac* operon is the example Step 1 likes to use to integrate these concepts

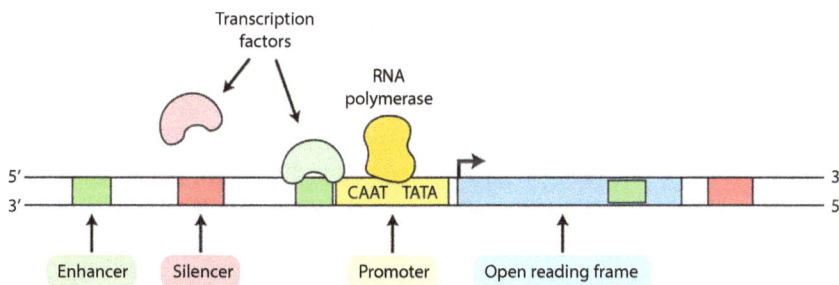

Figure 3-1. Regulation of gene expression.

Key sequences

Figure 3-2. The *lac* operon.

1. Set of co-expressed genes used by bacteria to activate lactose metabolism when lactose is available and glucose is not
2. Two regulatory steps must occur together to express the lac operon
   - **Low glucose**—activates CAP (binds enhancer)
   - **High lactose**—removes repressor protein (binds repressor)

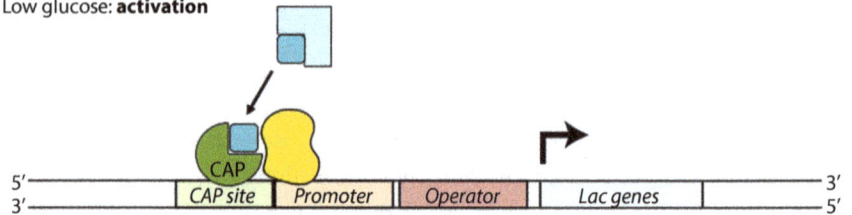

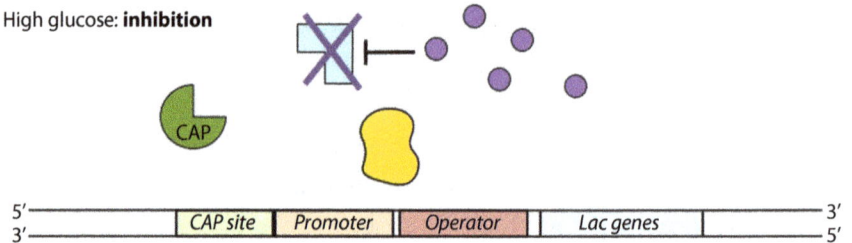

Figure 3-3. Glucose and the lac operon.

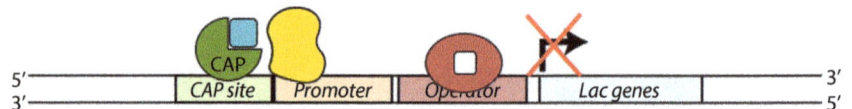

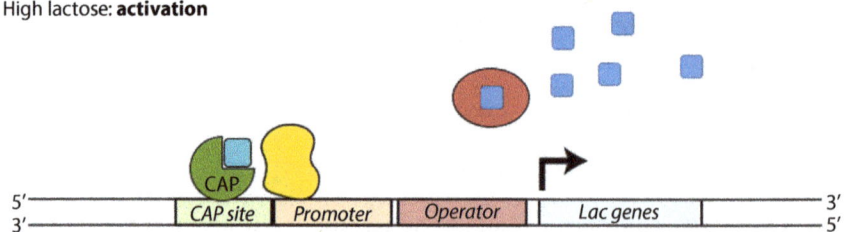

Figure 3-4. Lactose and the lac operon.

c. Mutations that disrupt gene regulation can be just as damaging as mutations in the gene's protein-coding region

   1. **Protein deficiencies**
      - In beta-thalassemia, mutations can occur in the promoter → less production of functional protein
      - Clinical: **Anemia, hemolysis**, weakness/fatigue
   2. **Increased activity**
      - Myc is a transcription factor that controls cell proliferation
      - Burkitt lymphoma, an aggressive form of cancer more common in children → caused by chromosomal translocation causing constitutive Myc expression
      - Clinical: Rapidly growing tumors on the jaw/face (endemic variant) or abdomen (sporadic); tumor lysis, hyperuricemia

3. Transcription

a. Four major types of RNA are found in the cell

   1. Ribosomal RNA (rRNA) is part of the ribosome, the main actor in translation
      - Synthesized by RNA polymerase I
   2. Messenger RNA (mRNA) encodes protein sequences
      - Synthesized by RNA polymerase II
      - Processed from heteronuclear RNA initially synthesized from DNA sequence in the nucleus
      - Amatoxins (poisonous mushrooms) inhibit RNA Pol II → apoptosis
   3. Transfer RNA (tRNA) links each mRNA codon into its corresponding amino acid
      - Synthesized by RNA polymerase III
   4. Micro RNA (miRNA) are short RNA sequences that do not encode protein
      - Regulates expression of proteins
      - Binds mRNA and prevents translation and/or tags for degradation

b. RNA is processed to become mRNA

   1. Poly-adenylation tails and 7-methylguanosine triphosphate caps are added to stabilize each end of the RNA molecule

Table 3-1. RNA subtypes and properties.

RNA type	Name	Function	Characteristics
rRNA	Ribosomal	Ribosome function	Most common
mRNA	Messenger	Encode protein	Largest RNA 5′ methylguanosine cap 3′ poly-A tail
tRNA	Transfer	Carry amino acids	Smallest Hairpin structure Unusual bases (Ψ, D)
miRNA	Micro	Regulation of gene expression	Short sequences Forms hairpin loops

Initial transcript: **hnRNA**

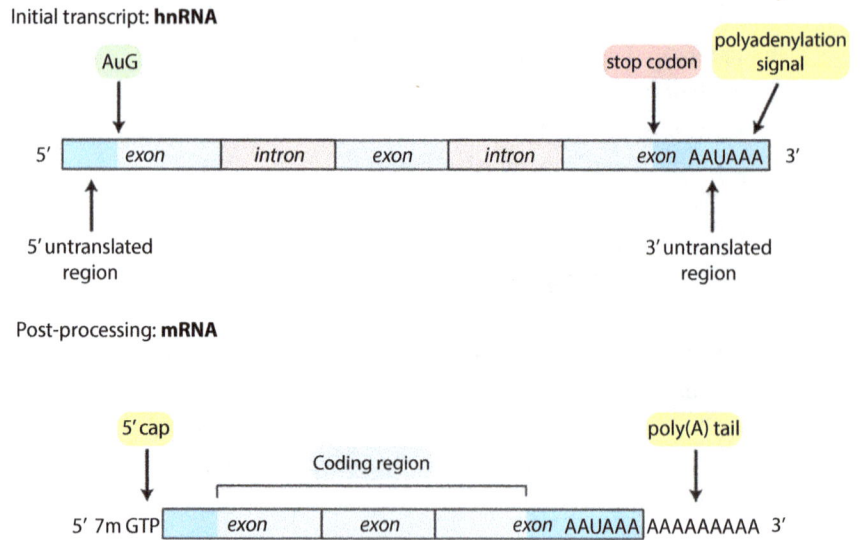

Figure 3-5. Processing of hnRNA into mRNA.

Initial transcript: **hnRNA**

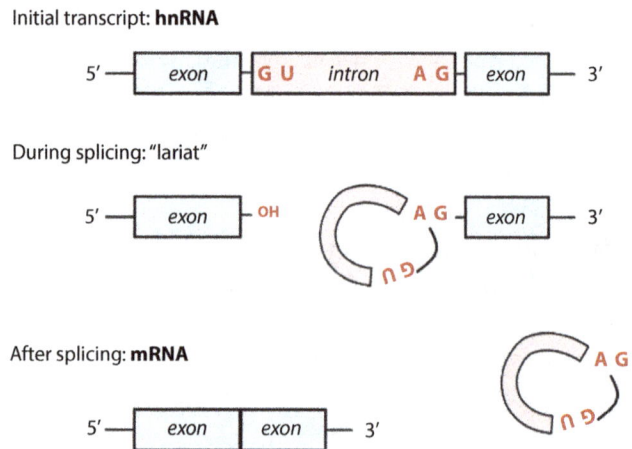

Figure 3-6. Splicing of introns.

2. Introns are removed through a splicing mechanism
   - Basic mechanism: Key sequences at each end of the intron are recognized and used to splice out the whole sequence
   - snRNPs are small ribonucleoproteins that "snip" out introns
   - Antibodies against splice proteins are specific for **systemic lupus erythematosus (SLE)**
   - Some introns are optional depending on the protein variant (e.g., membrane anchor vs. soluble)
3. All this processing happens in the nucleus, and mRNA is exported into the cytoplasm
   - Further processing is regulated in the cytoplasm by P bodies
4. miRNAs are small RNA molecules that bind to mRNA targets and inactivate/degrade them

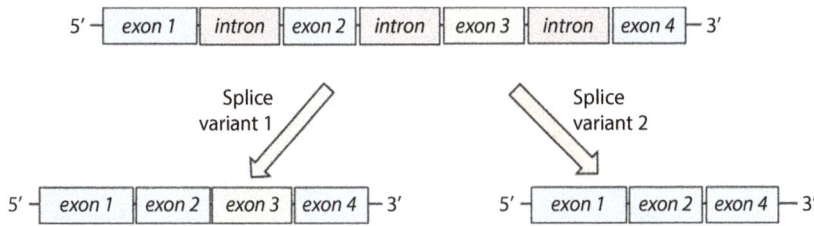

Figure 3-7. Differential splicing.

c. Clinical relevance
   1. Compounds that inhibit transcription can lead to severe symptoms because it prevents expression of many key proteins
   2. **Amatoxins** are the common example—from mushrooms
   3. There's another level of regulation at the RNA level, so conditions can mirror errors in DNA gene regulation
      • Loss of miRNA regulation → cancer

4. Translation

a. The mechanism of translation is particularly important for microbiology and pharmacology
b. mRNA is read as **codons:** Base triplets
   1. One codon (AUG) is the **start codon:** Initiate protein translation (and encode methionine)
   2. Three codons (UGA, UAA, UAG) are **stop codons:** Stop protein translation
   3. 61 codons (including AUG) encode amino acids
   4. The genetic code is **redundant**—multiple codons may correspond to 1 amino acid
c. tRNA helps translate the nucleic acid code into an amino acid sequence
   1. tRNAs have lots of unusual bases which can identify them vs. other RNA molecules
   2. Amino acids are covalently bonded to tRNA at the 3′ end, off the CCA sequence
      • Aminoacyl-tRNA synthetase — requires ATP

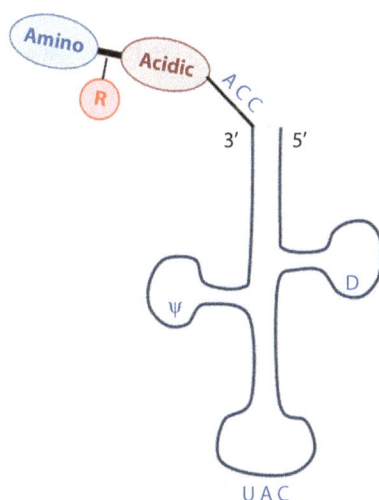

Figure 3-8. tRNA structure.

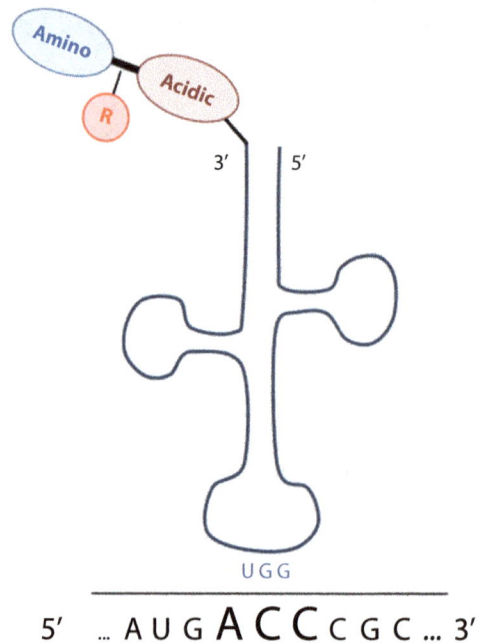

Figure 3-9. tRNA binding to mRNA.

3. The anticodon loop has 3 bases that bind to mRNA, "translating" each 3-base codon into the right amino acid
4. Wobble is a key feature of tRNAs that allows recognition of 2 or more codons encoding the same amino acid by a single tRNA
   - This allows all 61 codons encoding amino acids to be recognized by fewer than 61 different tRNAs
d. Ribosomes carry out translation
   1. Ribosomes contain rRNA and protein
      - Eukaryotic cells contain 40S and 60S subunits
      - Prokaryotic cells contain 30S and 50S subunits
   2. Assembly of the ribosome
      - The 40S subunit attaches to mRNA
      - A special initiator tRNA also binds and scans the mRNA for the start codon (AUG = methionine)
      - The 60S subunit attaches and the full ribosome is assembled
      - Bacterial toxins (EHEC, Shiga toxin, diphtheria toxin) inhibit ribosome assembly

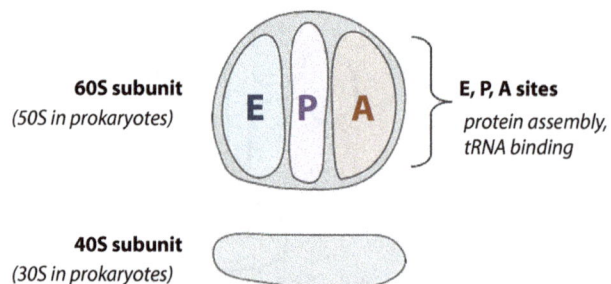

Figure 3-10. Ribosome structure.

3. Protein chain elongation
    - tRNA molecules loaded with specific amino acids enter the ribosome
    - The protein strand is transferred to the attached amino acid → elongation
    - The empty tRNA exits and the ribosome slides over
    - Signal recognition particle (SRP) associates and brings ribosome to the rough ER (if necessary)
        a. Recognized early in chain elongation
4. Termination
    - The stop codon (UGA, UAA, or UAG) indicates completion of protein synthesis and the new protein is released from the ribosome
5. Prokaryotes have 30S and 50S subunits in place of 40S and 60S, and the difference in structure allows for antimicrobial targeting
    - **30S inhibitors: Aminoglycosides** (e.g., neomycin) and **tetracyclines**
    - **50S inhibitors: Chloramphenicol,** macrolides (e.g., **erythromycin**), and **linezolid**

5. Protein folding

a. Proteins are synthesized as a simple strand → correct 3D structure is crucial for proper function
b. Three major steps of protein folding
    1. Folding of the strand can happen based on its amino acid sequence
    2. Modifications made to the sequence by other enzymes
    3. Chaperone proteins can stabilize proteins both as they are synthesized and transported
        - Some proteins require chaperone proteins throughout their lifespan

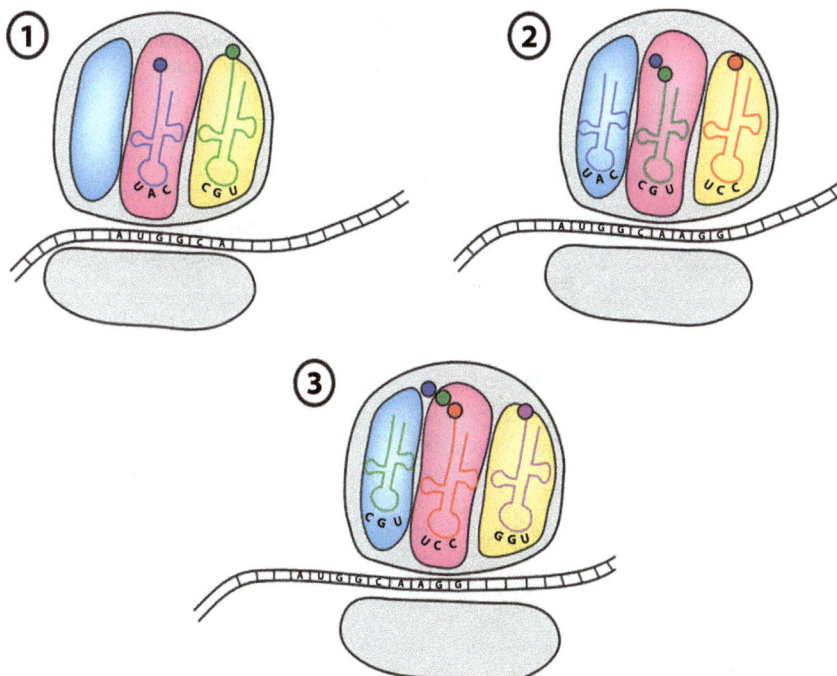

Figure 3-11. Protein synthesis in the ribosome.

c. Know the basic structure types and bonding types
1. **Secondary:** Hydrogen bonding into alpha helices and beta sheets
   - Alpha-helices are organized with all side groups facing out
     a. Transmembrane domains usually contain alpha-helices with mostly hydrophobic amino acids
     b. Proline (cyclical side group) and glycine (H side group) are unstable in alpha-helices due to angle of rotation
2. **Tertiary:** Disulfide bonds, hydrophobic/hydrophilic interactions
   - Cysteine residues form covalent bonds to stabilize folded structure
   - Active sites and regulation sites are important features of tertiary structures
3. **Quaternary:** Multiple subunits
   - Hemoglobin is a common example of quaternary structure
   - Hemoglobin is made of 4 subunits which work together to bind O_2
   - Sickle-cell anemia is a severe disease arising from a Glu → Val mutation in the β-globin subunit
     a. Deoxygenation leads to misfolding and aggregation of hemoglobin, "sickling" blood cells
     b. Acute phases (**crises**) with multiorgan pain, **autosplenectomy, hemolytic anemia;** more common in African Americans
4. The severity of a mutation in protein sequence usually correlates to how it affects protein structure
   - Hydrophobic ↔ hydrophilic
   - Specific amino acids (cysteine, proline)
   - Association between subunits

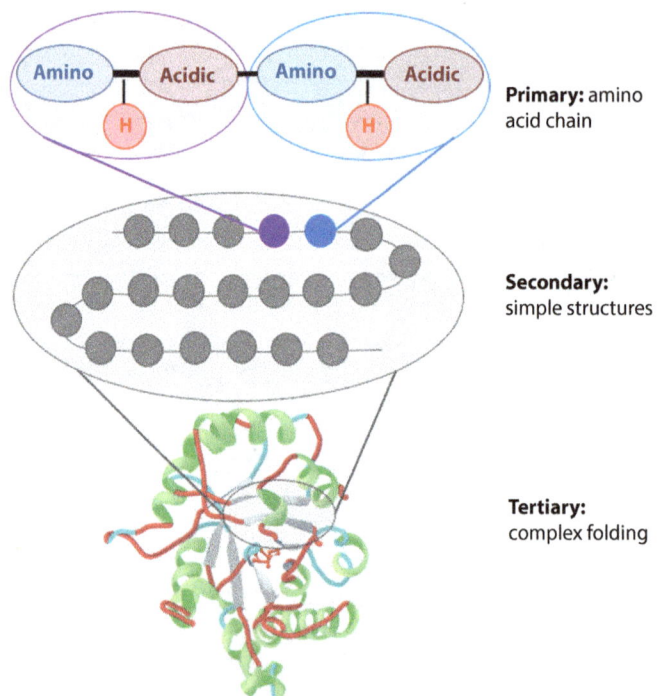

Figure 3-12. Levels of protein structure.

6. Post-translational modifications

a. Cleavage of protein strands enables correct folding/activity

   1. Proinsulin is activated by multiple cleavage steps, leaving inactive **C-peptide**

      - Injected insulin does not contain the C-peptide → measure C-peptide levels to assess endogenous insulin production by pancreatic β-cells

   2. POMC (proopiomelanocortin) is cleaved to β-endorphin (opioid), α-MSH (melanocytes), and **ACTH** (to regulate glucocorticoids)

      - **Addison's disease:** Loss of cortisol → increased ACTH → correlates with increased α-MSH

   3. Digestive enzymes largely exist as proenzymes in the pancreas to prevent aberrant activity

      - **Pancreatitis:** Alcohol use, drug reactions, etc. → cleavage (activation) of digestive enzymes → autodigestion of pancreas

b. Covalent modifications involve small molecules added to proteins to alter structure or affect activity

   1. Added by enzymes in the endoplasmic reticulum and Golgi

   2. Tags are added to direct proteins to specific locations

   3. Mannose-6-phosphate (M6P): Lysosomal targeting tag

      - **Inclusion-cell disease (I-cell):** Can't add M6P, so lysosomal enzymes are secreted to blood instead; **coarse facial features, developmental, delay, infections;** fatal in childhood

c. Additional functional groups can be added

   1. Acetylation, methylation, phosphorylation all common

   2. Stabilize, activate, deactivate the protein

d. Proteins are transported to their final location in vesicles

7. Collagen synthesis

a. One of the big biochem-specific topics tested on Step 1, for 2 reasons

   1. Ties together each step of post-translational modification

   2. Diseases result from mutations of different types across tissues

b. **Synthesis of collagen**

   1. The amino acid sequence has repeats of Glycine-X-X where X can be lysine or proline (think of these as "small" amino acids)

   2. In the ER, post-translational modification promotes collagen fiber formation into a triple helix

      - Hydroxylation adds –OH groups (think hydrogen bonding) (Vitamin C)

      - Glycosylation adds sugars (more O content added)

   3. Exported outside the cell, where it is cleaved at the ends, and crosslinked to create a meshwork extracellular matrix

c. **Types of collagen**

   1. Different parts of the body use different forms of collagen depending on the strength of the tissue and surrounding cell types

d. Clinical: Diseases can arise from errors in protein synthesis or in post-translational modification

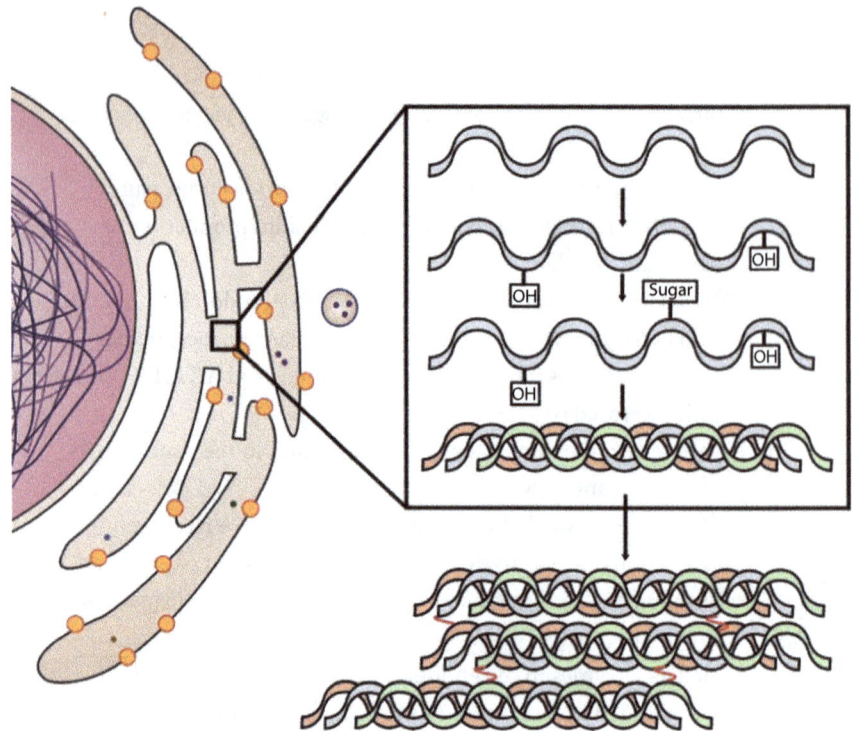

Figure 3-13. Collagen synthesis.

Table 3-2. Types of collagen.

Type	Location	Details/clinical correlates
I	**Bone** (and skin, tendons, ligaments, cornea, vessels)	90% of all collagen Osteogenesis imperfecta
II	Cartilage	
III	Blood vessels, intestines, lymphatics, bone marrow	Often associated with type I collagen Ehlers-Danlos **(vascular subtype)**
IV	Basement membrane	Alport syndrome
V	Associated with type I collagen	Ehlers-Danlos **(classic subtype)**

1. **Osteogenesis imperfecta** is caused by mutations of collagen in bone
   - Causes: Mutations in protein sequence affect folding and stability of the triple helix
     a. Leads to altered structure directly (e.g., bulky amino acid side chain)
     b. Inhibits normal post-translational modification (e.g., ER enzymes that hydroxylate proline)
   - Clinical: Severity depends on type of mutation
     c. **Brittle b**ones → hearing loss, tooth problems
     d. **Blue** sclerae → veins visible through matrix
2. **Ehlers-Danlos syndrome** is caused by impaired collagen processing
   - Causes: Mutations in processing of collagen
     a. Prevent cleaving tropocollagen after folding
     b. Prevent proper crosslinking due to altered post-translational modification

Table 3-3. Regulation of the *lac* operon.

Glucose level	Lactose level	*Lac* expression
High	High	
High	Low	
Low	High	
Low	Low	

- • Clinical: Many types depending on mutation and gene affected
  - a. Flexible joints and dislocations
  - b. Vascular weakness → **aneurysms**
  - c. Types III and V are most common
  - d. Can be autosomal dominant or recessive
  - 3. Vitamin C deficiency leads to impaired hydroxylation → decreased functional collagen (reversible)—bleeding from gums is classic presentation

8. Summary

- a. The **lac operon** is a common example of gene regulation by activators and repressors
- b. Transcription of mRNA involves synthesis and processing
  - 1. **Amatoxins** inhibit human RNA Pol II, leading to cellular apoptosis at sites of rapid protein synthesis (liver, GI, kidneys)
  - 2. Rifampin inhibits mRNA synthesis in TB
  - 3. miRNAs degrade or inhibit translation of specific mRNAs
- c. Protein synthesis inhibition is a key mechanism of both microbial toxicity and antimicrobial drugs because of specificity of prokaryote/eukaryote mechanisms
  - 1. *Diphtheria* and *Shigella*, and **EHEC** *E. coli* all inhibit protein synthesis
  - 2. 30S inhibitors: Aminoglycosides and tetracyclines
  - 3. 50S inhibitors: **Chloramphenicol, clindamycin, erythromycin**, and **linezolid**
- d. Protein folding and modification is integral to proper function
  - 1. **Hemoglobin** is tested as mutations that affect folding lead to sickle cell, thalassemia, etc.
  - 2. Scurvy: Defects in collagen folding due to vitamin C deficiency
  - 3. Osteogenesis imperfecta and Ehlers-Danlos: Defects in collagen synthesis leads to structural defects in bone and connective tissue

Practice questions

1. A cell biologist is studying the effect of a newly discovered cell surface protein on cellular development and differentiation. The researcher wants to induce gene transcription in an animal model by injecting a virus containing the gene of

interest, but wants to ensure that the gene is only expressed in a specific cell type. Which of the following sequences in the virally carried DNA molecule would effectively achieve this goal?

A. Enhancer
B. Intron
C. Poly-(A) signal
D. Promoter
E. Repressor

2. Researchers are studying how mutations in DNA affect protein expression in a newly identified genetic disease. They begin by quantifying RNA levels in cells from different patient populations. Which of the following mRNA sequences (5′ to 3′) corresponds to the open reading frame of this DNA?

5′ TATGATGATCGGATCGCTTGTTAACTGATAGAAT 3′
3′ ATACTACTAGCCTAGCGAACAATTGACAATCTTA 5′

A. UAUGAUGAUCGGAUCGCUUGUUAACUGAUAGAAU
B. UACUACUAGCCUAGCGAACAAUUGACU
C. ATGATCGGATCGCTTGTTAACTGATAGAAT
D. AUGAUGAUCGGAUCGCUUGUUAACUGA
E. AUGAUGAUCGGAUCGCUUGUUAACUGAUAG

3. A 17-year-old female comes to the free reproductive health clinic, but will not tell the triage nurse what her chief complaint is. When she sees the physician, she tearfully explains that she thinks she had unprotected intercourse and is worried she got a sexually transmitted infection. She has noticed milky discharge from her vagina and is in significant pain with any movement of her abdomen. The physician does an STI screen and diagnoses Chlamydia. She prescribes erythromycin, a protein synthesis inhibitor. Where does erythromycin act to exert its antibacterial effects?

A. 40S ribosome subunit
B. 50S ribosome subunit
C. Acetyltransferases in the cis-Golgi
D. RNA polymerase I
E. Signal recognition particle (SRP)

4. Researchers in an infectious disease laboratory are investigating the pathogenesis of the Shiga toxin of *Shigella dysenteriae* (the cause of dysentery). They have identified the ribosome as the toxin's site of action. Researchers note that inhibition of the ribosome leads to major decreases in protein synthesis. Which of the following properties of the ribosome make it unique as a catalytic molecule?

A. It has the ability to self-replicate.
B. It is highly conserved across all known living cellular organisms.
C. It is the largest protein complex found in eukaryotic cells.
D. Its catalytic elements are comprised of RNA.
E. Its coding sequence undergoes rapid mutation without loss of catalytic function.

5. A medical research group is recruiting individuals to participate in a new study to construct a transcriptome database, in which mRNA will be collected from several cell types and sequenced to determine which genes are expressed. Their hope is that a diverse collection of patient samples will help link disease to subtle alterations in gene expression. The first step in sequencing of mRNA is to reverse transcribe it to DNA. To ensure that only mRNA is sequenced, and not other types of RNA or DNA, what feature of mRNA should be targeted to initiate the reverse transcription process?

   A. 7-methylguanosine triphosphate
   B. Poly-adenylation signal
   C. Promoter regions
   D. Small nuclear RNA molecules
   E. Telomeres

6. A 38-year-old man presents to his primary care physician with breathlessness and difficulty swallowing that has been worsening over the past 2 years. He has also noticed wasting of the muscles distal to his knees. His father and sister suffer similar symptoms. The physician suspects a genetic cause, and after testing of all family members, a mutation in the desmin gene is identified. In each of the affected family members, alanine is replaced with proline in a highly conserved alpha-helical region of the protein. Unaffected family members do not have this mutation. What is the most likely cause of disease in these patients?

   A. Proline forms covalent bonds with nearby amino acids, creating a kink in alpha-helical structure
   B. Proline has a rigid structure that restricts proper alpha-helix formation
   C. Proline is hydrophobic, whereas alpha-helices typically contain hydrophilic amino acids
   D. Proline is too big to fit in the structure of an alpha-helix
   E. Proline's small size makes alpha-helical structure entropically unfavorable

7. A 13-year-old African American male suffers from sickle cell anemia. He has been repeatedly hospitalized for sickle cell crises and infections related to auto-splenectomy. He is asking the doctor to explain what is wrong with him. What is the most accurate explanation of the pathophysiology of his condition?

   A. A missense mutation in the active site of beta-hemoglobin prevents integration of heme into the 4 globin chains, impairing hemoglobin synthesis.
   B. A missense mutation in the hemoglobin beta-chain leads to an irreversible conformational change of hemoglobin in the deoxygenated state.
   C. A nonsense mutation in the hemoglobin beta-chain leads to decreased synthesis, altering the ratio of hemoglobin subunits and decreased functional hemoglobin tetramers.
   D. An activating mutation in the promoter of gamma-chain hemoglobin leads to increased levels of HbF, which binds tightly to oxygen and decreases delivery to tissues.
   E. An inactivating mutation in the porphyrin synthesis pathway leads to decreased heme production, impairing functional hemoglobin synthesis.

8. A 55-year-old man presents to a free clinic organized by students of the local medical school. He complains of a chronic cold, and notes that he also has had persistent pain in his gums. During the social history part of the interview, he explains that he is homeless and has inconsistent access to food, so he does not eat very well. If this patient suffers from vitamin C deficiency, what process is most likely directly impaired by this condition?

    A. Acetyl-CoA formation
    B. Crosslinking of polypeptide helices
    C. Glycosylation of hydroxylysine
    D. Proline hydroxylation
    E. Succinate dehydrogenation

9. A 21-year-old male comes in to the medicine clinic to see his physician for a regular checkup. He has a history of a disease affecting collagen synthesis, in which crosslinking of a particular type of collagen is impaired. He has been instructed to come to the physician regularly to track any new symptoms of the disease. Which of the following health effects would be most likely if this patient suffered a defect of collagen type III synthesis?

    A. Aneurysm
    B. Bone fracture
    C. Hyperextensibility of joints
    D. Rheumatoid arthritis
    E. Widespread shedding of epidermis

10. A 1-year-old boy is brought to the clinic because of poor motor function and failure to thrive. His facial structure looks slightly unusual as well. The physician suspects a problem with lysosome function. Samples of the patient's cells are examined by the pathologist, and she notices that the lysosomes have large inclusions. On the basis of this finding, the physician is able to make the diagnosis of I-cell disease. What is the most likely pathologic mechanism underlying this disease?

    A. Decreased acidity of lysosomes
    B. Decreased activity of lysosomal enzymes
    C. Decreased transcription of lysosomal enzymes
    D. Misfolding of lysosomal enzymes
    E. Mistargeting of lysosomal enzymes

CHAPTER 4

Functions of Proteins

1. Overview

  a. This chapter focuses on general concepts of protein function to lay groundwork for all of the metabolism pathways coming up

  b. There are some topics important for Step 1 that we focus on here

   1. Enzyme kinetics

   2. Muscle contraction

   3. Diseases: Marfan's syndrome, cystic fibrosis

  c. Try to group proteins into the ones that do something (i.e., enzymes) and the ones that don't do much (e.g., structural proteins)

  d. Learning about basics of protein function and nomenclature can help you make educated guesses on difficult questions

Structural proteins

  a. Some important diseases arise from defects in structural proteins, as opposed to proteins that carry out more active functions

2. Cytoskeletal proteins

  a. Intracellular structure and cell motility

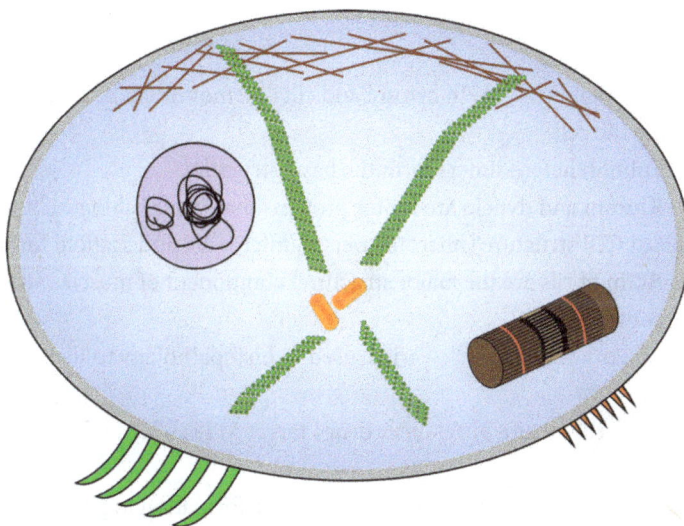

Figure 4-1. Cytoskeletal proteins.

Table 4-1. Intermediate filament stains (review).

Stain	Cell type	Used to identify
Cytokeratin	Epithelium	Epithelial tissue, e.g., skin, GI, ovarian, squamous cell carcinoma
Desmin	Muscle	Rhabdomyosarcoma, etc.
Vimentin	Mesenchymal tissue	Sarcomas, endometrial carcinoma, renal cell carcinoma, meningioma
Neurofilament	Neurons	Neuroblastoma
Glial fibrillary acidic protein (GFAP)	Glia (astrocytes)	Astrocytoma, glioblastoma

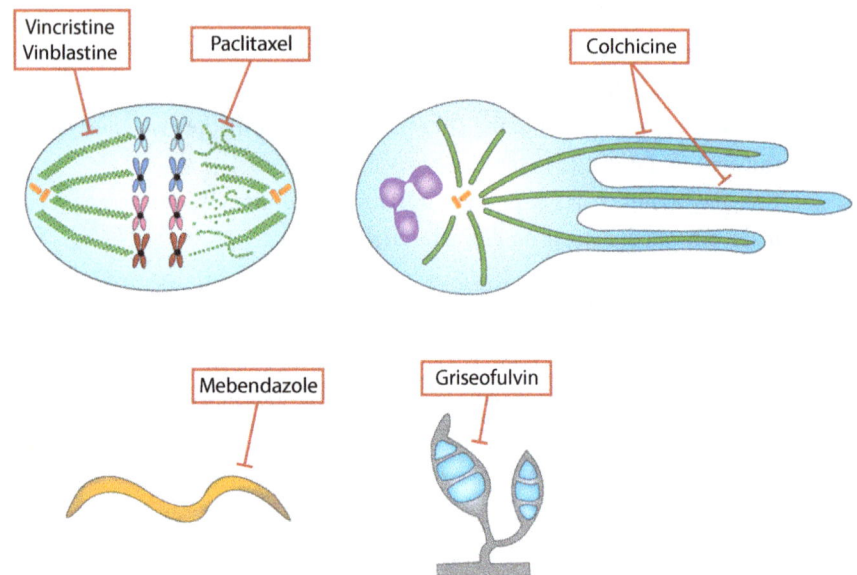

Figure 4-2. Anti-microtubule drugs.

b. **Microtubules (MTs):** In cytosol and cilia — movement of specific things inside the cell
1. Tubulin heterodimers form the basic structure
2. Kinesin and dynein are motor proteins that "walk" along MTs
c. **Actin:** Cell structure (microfilaments), microvilli, and scaffold for myosin
1. Actin fibrils are the major structural component of muscles and enable contraction
d. **Intermediate filaments:** Varied, used in histopathology to identify cell types
e. **Antimicrobial** and **anticancer** drugs target MTs as "active" in these cell types
1. **Vincristine** and **vinblastine** inhibit MT formation to prevent mitosis of cancer cells
2. **Paclitaxel** "locks" MTs in mitosis so cancer cells can't complete division
3. **Colchicine** affects neutrophil motility (important for **gout**)
4. **Mebendazole** and **griseofulvin** affect MT formation in parasites and fungi

    f. Microtubule defects lead to developmental disorders

       1. **Kartagener syndrome:** Cilia are dysfunctional because of MT motor protein → developmental, **reproductive**, **respiratory** problems

3. Extracellular matrix

    a. Extracellular space: Provides scaffold to support connective tissue and skin

    b. Collagen is a key component (reviewed in Chapter 3)

    c. Elastin: Springy substance in organs that need to stretch

       1. Like collagen, has glycine, proline, lysine

       2. Also like collagen, crosslinking of fibers requires lysyl oxidase

         • Inhibition of lysyl oxidase can cause aortic weakness (mimicking Marfan's syndrome)

       3. Differs from collagen on a biochemical level: No hydroxylation

    d. Fibrillin: Another elastic protein that complexes with elastin to provide stretch

    e. α_1-antitrypsin deficiency is due to excessive breakdown of elastin because of overactive elastase → **emphysema**

       1. Lungs primarily affected: Panacinar emphysema

       2. Liver also affected: Site of α_1-antitrypsin synthesis

         • Liver is damaged due to accumulation of misfolded protein

    f. Marfan's syndrome is due to mutation in fibrillin, preventing proper elastin function

       1. **Autosomal dominant** with variable expressivity (not all patients show same combination of symptoms—see Chapter 10)

       2. Tall stature — elastin doesn't keep them "pulled down"

       3. Eye problems — lens needs to be stretchable

       4. Heart problems (aortic dissection, valve problems) — large vessels need to be elastic

Active proteins

    a. Besides their role in structural support, proteins can have a number of more "active" functions; some of the major groups are

       1. Channels and pumps—allow flow of substances across membranes

       2. Receptors—receive signals from outside the cell

       3. Enzymes—facilitate chemical reactions

    b. Channel proteins are important in cardio and renal (antiarrhythmics and diuretics)

    c. Receptors are important in psychiatry, endocrine, and cancer

    d. Enzymes are important throughout, including in metabolism

4. Channel proteins

    a. Ionic and hydrophilic compounds cannot cross plasma membranes

    b. Compounds cross membranes through (1) passive diffusion, (2) facilitated diffusion, and (3) active transport

    c. **Simple diffusion** is when small molecules pass through the plasma membrane directly

       1. Gases (e.g., O_2)

       2. Lipids (e.g., steroids)

       3. Small, uncharged molecules (e.g., water in small amounts)

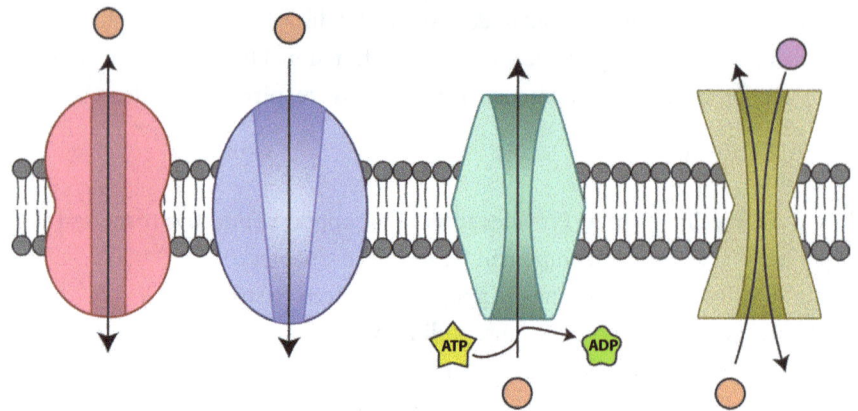

Figure 4-3. Membrane transport.

    d. **Facilitated diffusion** allows molecules (often ions) to flow down their electrochemical gradient through membrane proteins
      1. No energy is required
      2. Many neurotransmitter receptors are ion channels that allow free flow of specific ions
      3. Gap junctions help cells coordinate their activity (e.g., cardiac contraction)
    e. **Active transport** allows movement of compounds against their electrochemical gradients
      1. Energy is required: Using ATP, or coupling to energetic reaction
      2. Sodium-potassium ATPase: Common example, antiarrhythmics
        • Energy is required to move sodium out and potassium in
        • In resting conformation, Na^+ moves into the channel complex
        • ATP phosphorylation of the channel changes conformation and releases Na^+, binds external K^+
        • Release of P_i moves channel back to resting state, bringing K^+ in
    f. **Clinical:** Manipulating channel proteins is important in Cardiology, Neurology, and Renal sections
      1. Cardiac glycosides such as **ouabain**, **digoxin**, and **digitoxin** increase cardiac contractility by inhibiting Na^+-K^+-ATPase → inhibit Na^+/Ca^{2+} exchange → more Ca^{2+} inside the cell
      2. Action potentials rely on voltage-gated ion channels, inhibited by **tetrodotoxin** from pufferfish
      3. Diuretics in renal pharmacology manipulate ion channels to alter water resorption
        • Loop diuretics such as **furosemide** inhibit $Na^+/K^+/2Cl^-$ cotransporter
    g. Cystic fibrosis results from a defective Cl^- channel (CFTR)
      1. Decreased Cl^- transport impairs the co-secretion of Na^+ and water → thick secretions
      2. **Salty sweat** due to impaired reabsorption of Cl^- and Na^+ during sweat production
      3. Clog lungs, pancreas, bile ducts
      4. **Lung infections** are common, and often cause of death

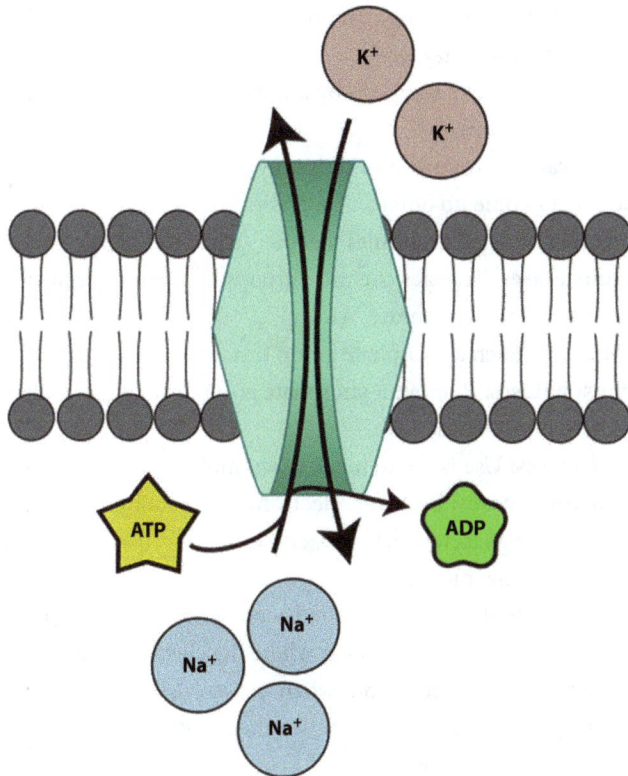

Figure 4-4. Sodium-potassium ATPase.

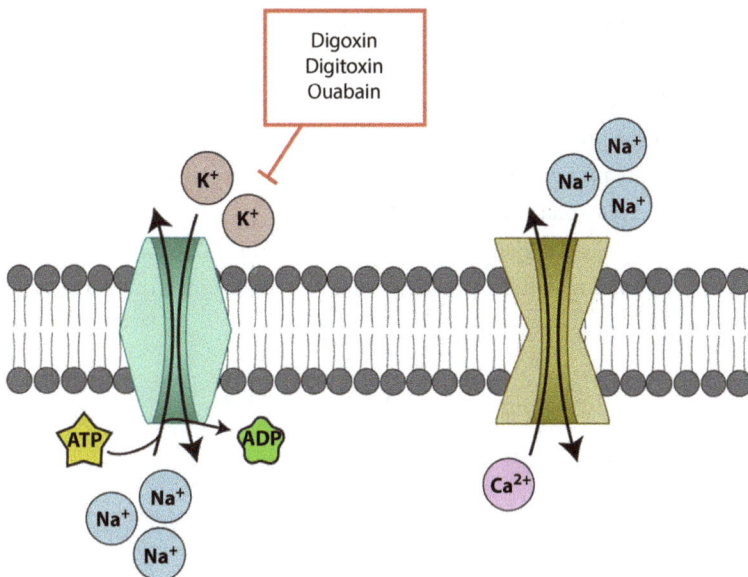

Figure 4-5. Cardiac glycosides.

5. **Malabsorption** as pancreas starts to fail
6. Notice **meconium ileus** in newborns as early sign

5. Enzymes

a. Enzymes are relevant to every organ system and play a major role in normal and disease states
b. An enzyme is a protein that facilitates a chemical reaction

1. Many chemical reactions cannot take place in living organisms without assistance from enzymes
2. Enzymes facilitate reactions by lowering the energy requirement to get the reaction started

c. Enzymes can be grouped into classes by their actions
d. Some classes come up often and are worth recognizing
   1. **Ligases:** Attach 2 molecules together (e.g., DNA ligase)
   2. **Transferases:** Transfer chemical groups between molecules (e.g., alanine amino transferase, ALT)
   3. **Kinases:** Attach a phosphate group to a molecule (e.g., protein kinase A)
   4. **Phosphatases:** Remove a phosphate group from a molecule (e.g., protein phosphatase 1)
   5. **Hydrolases:** Use water to break down molecules (e.g., protease)
   6. **Dehydrogenases:** Exchange electrons in reduction-oxidation (redox) reactions (e.g., lactate dehydrogenase)

e. Regulation of enzyme activity
   1. Enzyme activity can be regulated acutely, or over a long period of time
   2. The active site of an enzyme is where the chemical reaction occurs
   3. An active site only functions when in a very specific conformation
   4. Altering the structure of the active site is a straightforward way to regulate enzyme activity
      - **Production/Degradation:** Regulate total available levels of enzyme over a long period of time (hours)
      - **Activation/Deactivation by cleavage:** Altering existing pools of enzyme by cleaving off a sequence irreversibly allows the body to rapidly activate or inactivate (e.g., digestive enzymes)
      - **Activation/Deactivation by phosphorylation:** Addition or removal of the phosphate group → major regulator of enzyme activity and easily reversible with the help of external protein regulators (kinases and phosphatases)
      - Agonists/Antagonists: Small molecules can also regulate enzyme activity by binding to the active site or other regulatory sites, usually reversibly

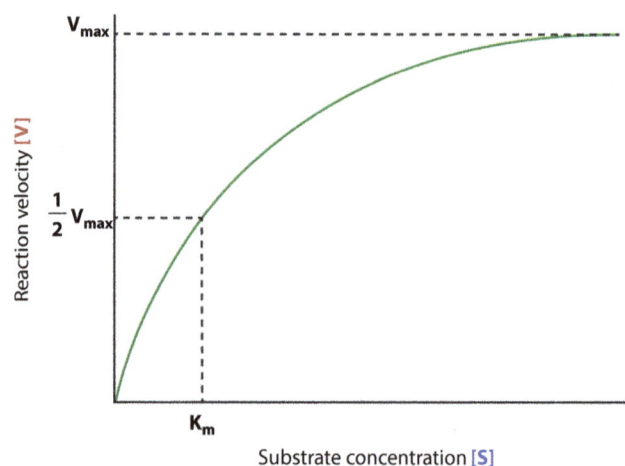

Figure 4-6. Enzyme kinetics.

5. Regulation of enzymes can get very complex, but these basic concepts come up over and over again

6. Many of these principles also apply to proteins in other contexts, such as regulation of AMPA receptors at synapses

6. Enzyme kinetics

a. Step 1 tests many metabolic pathways that involve multiple enzymes
   1. When multiple reactions chain together, kinetics becomes important (e.g., rate-limiting step)
   2. Understanding kinetics can help you reason through questions you may not have known otherwise

b. Basic concepts
   1. Every enzyme-facilitated reaction can be expressed in equation form

   $$[Enzyme] + [Substrate] \leftrightarrow [E\text{-}S] \leftrightarrow [Enzyme] + [Product]$$

   2. Movement of molecules is random, so enzymes rely on running into substrates and latching on with high specificity
   3. Concentrations of enzyme, substrate, and inhibitors directly affect the reaction rate
   4. Step 1 tests kinetics using the Michaelis-Menten model: Graph enzyme activity as **reaction velocity (V)** vs. **substrate concentration [S]**
   5. V_{max} = Highest possible reaction velocity an enzyme can achieve
   6. K_m = Substrate concentration [S] that allows the enzyme to operate at ½ V_{max}

c. Every enzyme in a pathway can have a different velocity, but the pathway as a whole is limited to the lowest velocity → rate-limiting step

d. Enzyme inhibition: Many toxins and drugs operate by inhibiting enzyme activity
   1. Competitive inhibitors look like substrate, and bind active site to prevent substrate binding
      - Names are also often similar (because chemical structures are similar)
      - Because substrate can displace inhibitor, adding a lot more substrate can overcome inhibition
      - **6-mercaptopurine** inhibits de novo purine synthesis
      - **Ethanol** treatment of methanol intoxication

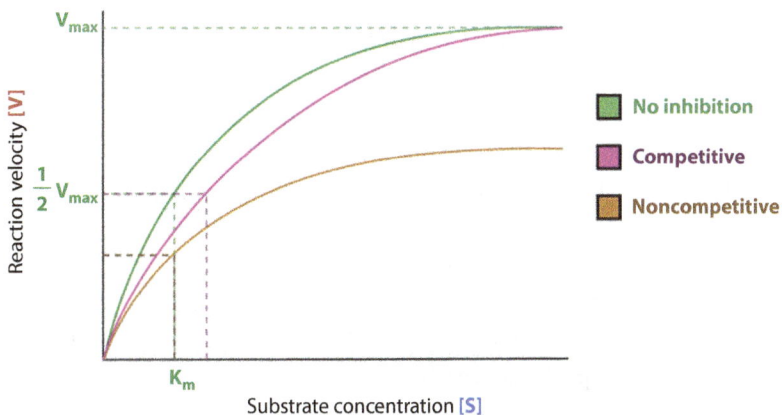

Figure 4-7. Enzyme inhibition.

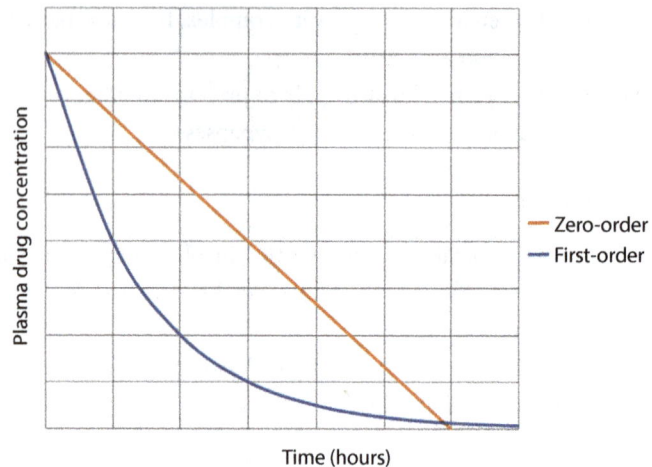

Figure 4-8. Drug elimination kinetics.

2. Noncompetitive inhibitors bind the enzyme somewhere else and alter conformation to make it harder to bind substrate
   - No amount of substrate can completely overcome inhibition as the enzyme is always structurally altered
   - **Allopurinol** ("allosteric") in gout
   - **Nifedipine**, an antihypertensive blocking Ca^{2+} channels
3. Irreversible inhibitors covalently bind and inactivate enzymes
   - Essentially reduce [E]: Decrease V_{max}, but K_m stays the same
   - **Aspirin** — anti-inflammatory (inhibits COX enzymes)
   - **Organophosphates** — insecticides (inhibit acetylcholinesterase)
e. There are 2 levels of knowing enzyme kinetics: (1) Understanding trends and (2) calculating specifics
f. Reaction kinetics is tested as part of dosage calculations in pharmacology
g. Most relevant to enzyme kinetics is reaction order: Product vs. time
   1. **Zero-order** reaction velocity (slope) doesn't depend on [S]
   2. **First-order** reaction velocity is proportional to [S]
   3. **Second-order** reactions are proportional to [S]² (the square of substrate concentration)

7. **Signal transduction**

   a. Cells have to be able to respond to external signals
   b. There are several pathways the cell uses, depending on the type of signal molecule
      1. G-protein signaling—cell surface receptors
      2. Tyrosine-kinase signaling—cell surface receptors, usually for growth factors
      3. Nuclear signals—for thyroid and steroid hormones
      4. Signal-responsive channels—for example, neurotransmitter receptors such as acetylcholine (ACh) receptors
   c. G-protein coupled receptors (GPCRs)
      1. All GPCRs have similar properties

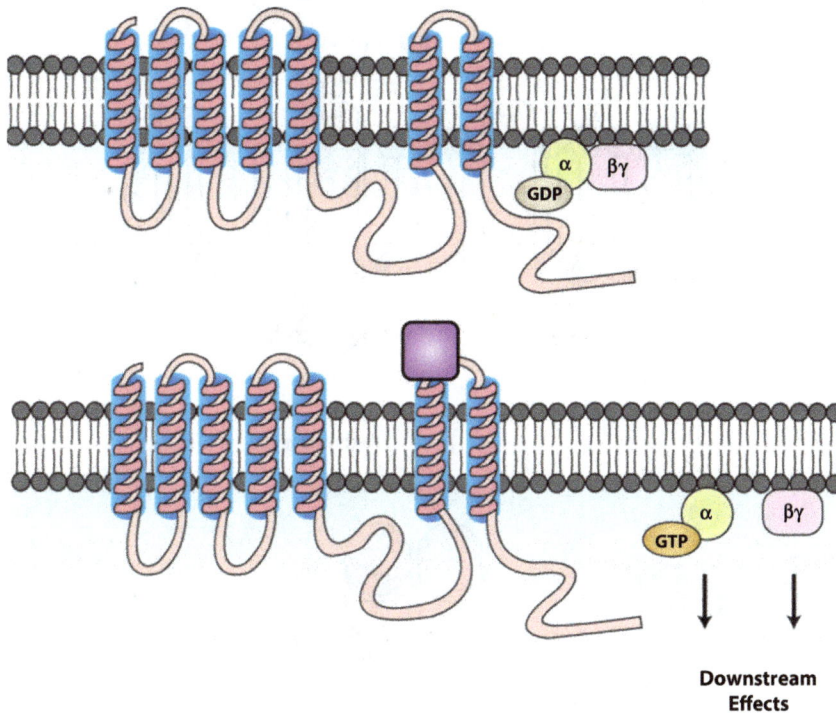

Figure 4-9. G-protein coupled receptors.

Table 4-2. G-protein coupled receptor pathways.

G_α type	Signaling pathways	Mechanism
G_s	Epinephrine (β_1, β_2, β_3) Dopamine (D_1) Histamine (H_2) Vasopressin (V_2)	Adenylyl cyclase → cyclic AMP (cAMP) → protein kinase A
G_q	Epinephrine (α_1) Histamine (H_1) Muscarinic (M_1, M_3) Vasopressin (V_1)	Phospholipase C → (PIP_2 → DAG, IP_3) → protein kinase C, $[Ca^{2+}]_i$
G_i	Epinephrine (α_2) Muscarinic (M_2) Dopamine (D_2)	**Inhibit** adenylyl cyclase → ↓cAMP

- Membrane protein with 7 transmembrane domains
- Signaling molecule binds to protein outside the cell → activate intracellular G-protein subunits
- Binding of GTP activates G_α subunit

2. GPCR pathways depend on what the G_α subunit does upon activation

d. Receptor tyrosine-kinases (RTKs)

   1. This pathway is commonly used by growth factors, so mutations may lead to cancer

   2. Two key facts to know about RTKs
      - When activated by a signaling molecule on the outside, they autophosphorylate on the inside

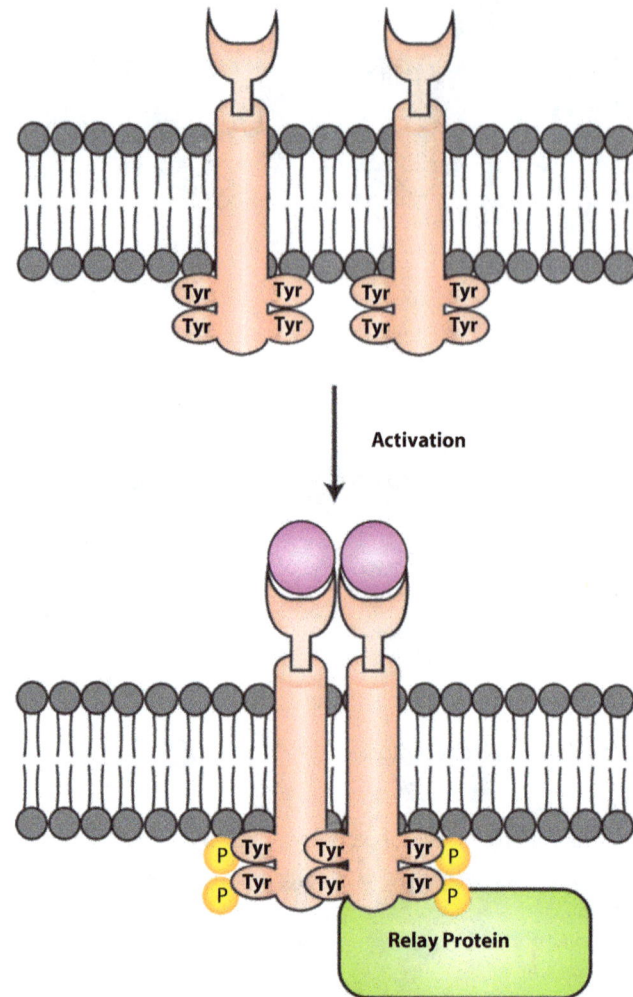

Figure 4-10. Receptor tyrosine kinase.

- They signal through **Ras**, a small GTPase which amplifies the signal and is common to signaling pathways in response to many growth factors
- Mutations that lead to constitutive activity of either are oncogenic: A single mutated copy is enough to significantly increase cell division
  e. Intracellular receptors
    1. Lipophilic compounds: Thyroid hormone, steroid hormones
       - **Steroid hormones** have receptors in cytosol and translocate to nucleus
       - **Thyroid hormone** moves directly to the nucleus, where the receptor is
    2. In both cases, they bind to transcription factors to affect gene expression
    3. Signal cascade is much slower than extracellular signal cascades, though effects can last longer

8. **Muscle contraction**

   a. Contraction of muscle integrates the biochemical concepts of (1) matrix proteins, (2) motor proteins, and (3) signal transduction
   b. The sarcomere is the smallest unit of muscle contraction
      1. Actin forms the scaffold (thin filament; I band)

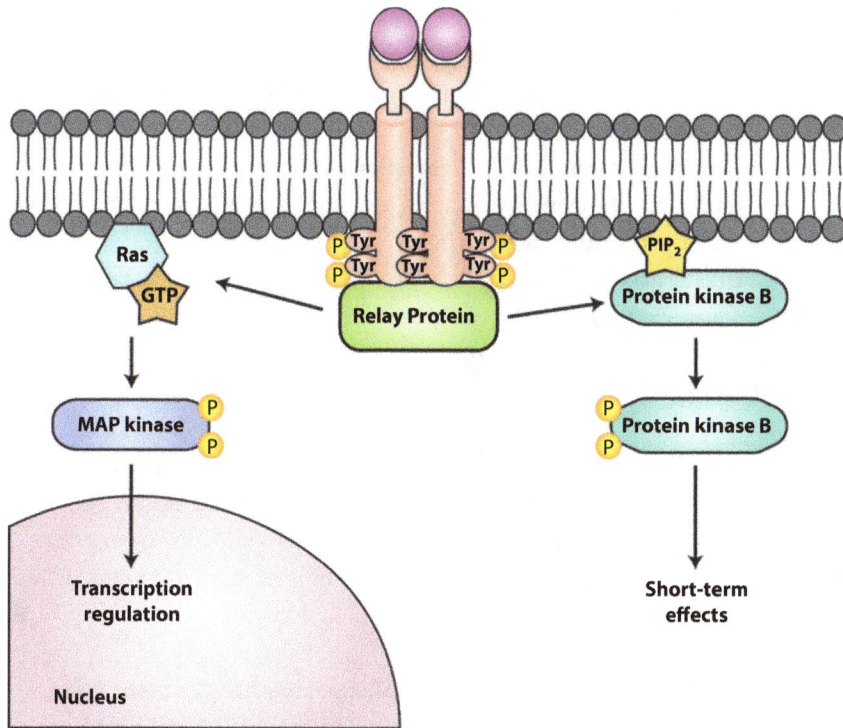

Figure 4-11. Insulin receptor pathways.

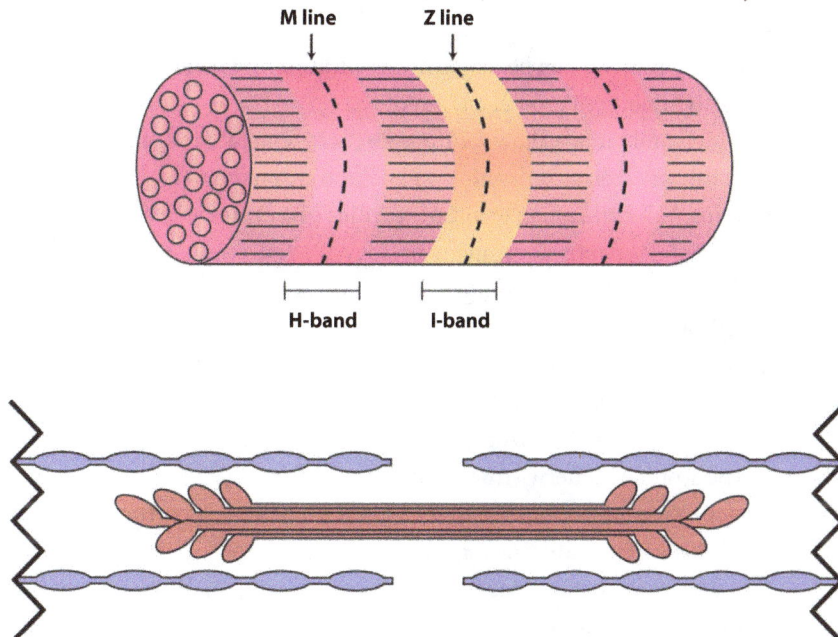

Figure 4-12. Sarcomere and muscle contraction.

2. Myosin motor protein bundles (thick filament; H band) walk along actin to squeeze (contract) muscle (A band: overlap)

3. ATP is required to release myosin and start a new contraction cycle

   • **Rigor mortis** occurs when ATP is lost: Muscle fibers contract with ADP, but can't release without ATP

c. Muscle contraction relies on Ca^{2+} signaling
1. Binding of neurotransmitter (e.g., acetylcholine) to ion channels induces conformational change → activates channel (allows flow of ions into cytosol)
2. Coupled to opening of Ca^{2+} channels on sarcoplasmic reticulum
3. Ca^{2+} binds the troponin complex, releasing tropomysin from actin and allowing myosin to bind actin
4. During acute **myocardial infarction (MI)**, dying cardiac muscle cells release **troponin**, which can be measured in plasma

d. **Dystrophin** anchors actin cytoskeleton and intracellular structures (including sarcomeres) to the cell membrane, allowing coordinated contraction

e. **Muscular dystrophy:** Errors in dystrophin (X-linked), loss of muscle anchoring → progressive muscle damage and eventual loss
1. Progressive muscle weakness, starting in pelvis and moving outward
2. Dilated cardiomyopathy resulting from progressive muscle destruction
3. **Duchenne:** Deletion of dystrophin (frameshift mutation)—onset typically <5 years old
4. **Becker:** Better prognosis (partial deletions)—onset in adolescence

f. **Myotonic dystrophy:** Myotonin protein kinase mutated (AD), leads to muscle wasting, arrhythmias, but also **myotonia** (sustained contraction)

9. **Summary**

a. Proteins are central to biological activity, whether through structural support or activity

b. Structural proteins can be targeted with **antimicrobial** and **anticancer** drugs to inhibit cell growth and motility

c. Channels and pumps allow for transport across membranes and play a large role in cardiac contraction and neurotransmission
1. Drugs affecting ion flow include **cardiac glycosides** and are used in cardiology, nephrology, and neurology
2. **Cystic fibrosis** is a severe genetic disease resulting from a defect in the CFTR ion channel

d. Enzymes play a role in many chemical reactions, and are often named for their primary function
1. Interpreting the name of an enzyme can help reason through a question you wouldn't otherwise be able to answer

e. Enzyme kinetics are important in metabolism to help understand concepts of **inhibition** and **rate-limiting steps**
1. Rather than memorizing specific equations, focus on general changes in different situations
2. Competitive inhibitors compete for binding in the active site
3. Noncompetitive inhibitors bind an allosteric site
4. Irreversible inhibitors permanently inactivate an enzyme

f. Signal transduction pathways play important roles in many situations and often allow for amplification of various downstream pathways
1. **GPCRs** function in a wide range of signaling systems
   - **Pseudohypoparathyroidism** is caused by a mutation in the G_s subunit of the PTH receptor

- Hypocalcemia despite increased PTH implies signaling problem
- Pattern of inheritance modified by **imprinting** (paternal copy mostly silenced)

2. **RTKs** function in growth hormone pathways, and dysfunction can lead to cancer
   - HER2/neu is mutated in 30% of breast cancer cases
   - Mutation leads to constitutive activity → unrestrained cell growth

g. Muscle contraction relies on myosin motor protein walking along anchored actin

1. Coordination of extracellular signals releases Ca^{2+} that spreads all along the muscle cell
2. **Troponin**, a protein involved in contraction, is used as a marker for myocardial infarction (MI)
3. Muscle relaxation relies on ATP
4. Defects in muscle fiber anchoring → muscular dystrophy with muscle wasting

Practice questions

1. A medical student is working in a basic science laboratory that studies diseases that affect muscle function. They use techniques such as electron microscopy to study how mutations affect the molecular structure of muscle and correlate this with changes in muscle contraction. The student examines several samples of muscle imaged using electron microscopy and notes small differences in the striated muscle organization. In particular, in one sample he notices that, relative to control samples, there is a smaller change in the width of the H band in contracted muscle. What does this observation suggest?

A. Less tropomyosin binds to actin
B. Movement of myosin along actin decreases
C. Myosin has lower binding affinity to thick filaments
D. There is also a smaller change in A band width
E. There is no change in width between Z lines

2. A 70-year-old woman is killed when her car collides with another car. As part of the investigation into the cause of the crash, the police hope to get an autopsy report to determine how the woman died. The pathologist is unable to begin the autopsy for several hours; however, because by the time the woman is brought to him, her muscles have grown extremely stiff with rigor mortis. What is the molecular explanation for this finding?

A. Actin filaments do not release actomyosin without ATP
B. Muscle cells have enough ATP to sustain active contraction after death
C. Myosin filaments do not bind actin without ATP
D. Myosin filaments do not release actin without ATP
E. Sarcomeres are stretched without ATP until sufficient protein degradation occurs

3. An 18-year-old male who is the star player on his school's basketball team collapses on the court during a regular season game. He cannot be revived, and an ambulance is called. He is rushed to the hospital but is pronounced dead on arrival. An autopsy reveals that the cause of death was aortic dissection. It

appears that the patient suffered a genetic disease that predisposed him to this condition, although he never experienced pain or other symptoms that led him to seek treatment. Which protein was most likely mutated in this patient?

A. Axonemal dynein
B. Fibrillin
C. Lysyl hydroxylase
D. Type I collagen
E. Type III collagen

4. A neurology fellow is working in a laboratory that studies action potentials and membrane conductance and their relation to psychiatric disease. He is focusing on how genetic mutations can affect intracellular ion concentrations. In neurons derived from normal controls, the membrane potential relative to extracellular fluid is –70mV. In neurons cultured from patients, however, the membrane potential is only –40mV. Which of the following most likely accounts for this observation?

A. Active transport
B. Endocytosis
C. Electron transport chain
D. Passive diffusion through channel proteins
E. Passive diffusion through the membrane

5. A 52-year-old woman presents to her gynecologist complaining of hot flashes and night sweats, which has led to significant insomnia. She asks the physician to prescribe her hormone replacement therapy, as she has heard that this significantly reduces symptoms. After taking a history and physical exam, the gynecologist agrees she is a good candidate and begins her on an estrogen transdermal patch. Where does this hormone bind to its receptor at the target cell?

A. Cell surface
B. Cytosol
C. Endoplasmic reticulum
D. Mitochondria
E. Nucleus

6. A 3-year-old boy is brought to the neurology clinic because his parents are worried he may have epilepsy. They describe several episodes of muscle spasms and twitching. Upon physical exam, the physician also notes abnormally short stature and poor bone development, which combined with the neurologic symptoms, leads the physician to suspect chronically low calcium secondary to poor parathyroid function. Indeed, a blood test shows hypocalcemia and hyperphosphatemia. Unexpectedly, however, levels of parathyroid hormone are elevated. The physician suspects a mutation in the G-protein coupled receptor (GPCR) system that affects adenylyl cyclase (AC). Which combination of findings is most likely found in this patient?

A. Decreased G protein activation, decreased AC activation
B. Decreased G protein activation, increased AC activation
C. Increased G protein activation, decreased AC activation
D. Increased G protein activation, increased AC activation
E. No change in G protein activity, decreased AC activation
F. No change in G protein activity, increased AC activation

7. An 82-year-old woman is referred to the oncologist's clinic after referral from her primary care physician. A recent mammogram showed a possible mass, and a biopsy leads to a diagnosis of invasive ductal carcinoma. Unfortunately, a PET scan reveals multiple metastases to the bones, making recovery unlikely. A pathologist involved in her treatment asks for consent to investigate the molecular properties of her cancer, and she consents. Genetic analysis reveals multiple mutations, including an activating mutation of heparanase, a hydrolase involved in basement membrane breakdown prior to metastasis. What is the most likely function of this enzyme?

    A. It catalyzes alterations in the conformation of existing molecules.
    B. It catalyzes cleavage of covalent bonds with the aid of a water molecule.
    C. It catalyzes formation of double bonds as a method of splitting a molecule into smaller parts.
    D. It catalyzes formation of new C–O bonds to join 2 molecules together.
    E. It catalyzes transfer of a small carbon functional group from one molecule to another.

8. A 20-year-old patient suffers from glucose-6-phosphate dehydrogenase (G6PD) deficiency, which results in a severe hemolytic anemia syndrome. The patient is sent to a geneticist to assess the cause of the disease and identify potential treatments. Genetic analysis reveals a newly discovered mutation in the coding region of G6PD. This mutation leads to a large increase in the Michaelis constant (K_m) of the reaction. What does this indicate about enzyme function?

    A. A larger quantity of enzyme must be synthesized to approach the maximum velocity (V_{max}) of the reaction in a healthy patient's cells.
    B. The equilibrium constant K_{eq} of the reaction decreases.
    C. The mutated enzyme can still catalyze the dehydrogenation of glucose-6-phosphate, but it is unable to release the product into solution.
    D. The mutated enzyme has a higher affinity for glucose-6-phosphate.
    E. The mutated enzyme has a lower affinity for glucose-6-phosphate.

9. A 45-year-old man with a history of high blood pressure reports to the ER with chest pain. An electrocardiogram (ECG) is performed and, based on the results, the physician suspects a non-ST-elevation myocardial infarction (NSTEMI). A follow-up test is run to quantify levels of troponin. Which of the following statements is most accurate regarding troponin?

    A. Myocardial infarction leads to decreased plasma levels of troponin due to inhibited cardiac muscle contraction.
    B. Troponin binds myosin, facilitating myosin movement along actin filaments in the sarcomere.
    C. Troponin is involved in the programmed cell death cascade (apoptosis), making it valuable for identifying infarction in many tissue types.
    D. Troponin is normally involved in muscle contraction but is released by dying cells after myocardial infarction.
    E. Use of the test has greatly decreased in recent years because it requires invasive biopsy of the heart tissue.

CHAPTER 5

Glucose Metabolism

1. Purpose

a. This chapter focuses on the main pathways and enzymes that regulate processing of glucose into energy

b. Think about it as a highway from glucose to ATP, with many entrances and exits to different pathways depending on what the cell needs

c. Step 1 also likes to ask about how fed/fasting state affects the balance of different pathways, so understanding how pathways connect is key

2. Overview of metabolic pathways

a. On the "highway" of glucose to ATP, the central pathways are the focus of this chapter

b. Pathways can be generally categorized as **"fed-state"** or **"fasting-state"** based on whether they build energy reserves or release energy for later use

c. We can further break down the processes into 4 main categories

    1. Breakdown of energy sources: Glycolysis; fructose/galactose metabolism; the tricarboxylic acid (TCA) cycle; and oxidative phosphorylation

    2. Storage of energy: Glycogenesis and lipid synthesis

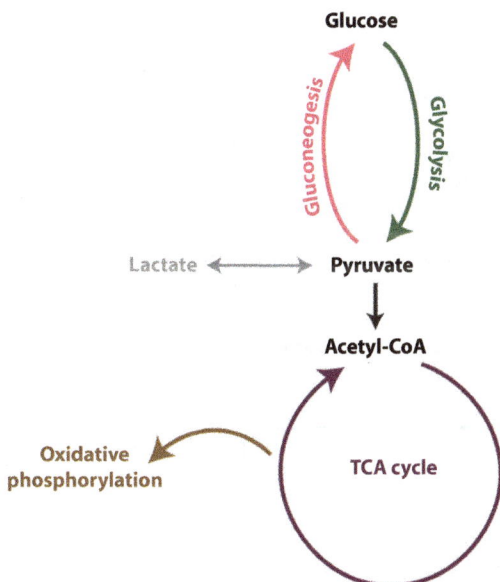

Figure 5-1. Overview of metabolic pathways.

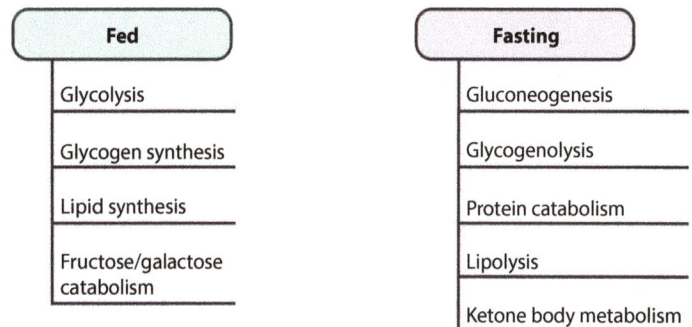

Fed
Glycolysis
Glycogen synthesis
Lipid synthesis
Fructose/galactose catabolism

Fasting
Gluconeogenesis
Glycogenolysis
Protein catabolism
Lipolysis
Ketone body metabolism

Figure 5-2. Pathways in fed and fasting states.

3. Fasting state metabolism: Gluconeogenesis, lipolysis, glycogenolysis, protein catabolism

4. Side pathways (other functions): Pentose phosphate pathway (HMP shunt), urea cycle, cholesterol synthesis

d. Side pathways contribute energy from other sources and/or create energy stores and will be discussed in subsequent chapters

3. Breakdown of glucose

a. Glucose may enter the cell from outside or be released from intracellular stores (glycogen)

b. Glycolysis occurs in the cytosol

1. Glucose is broken down into smaller components

2. Only 2 molecules of ATP per glucose are produced

3. Other carbohydrates, glycogen, and lipid metabolites (glycerol) enter the glycolysis pathway

c. The TCA cycle takes over in the mitochondria when oxygen is present

1. This produces the intermediates NADH (derived from vitamin B_2) and $FADH_2$ (derived from vitamin B_3)

2. Amino acids can also enter into the TCA cycle to supply energy

d. Oxidative phosphorylation occurs on the inner mitochondrial membrane to turn NADH and $FADH_2$ into ATP

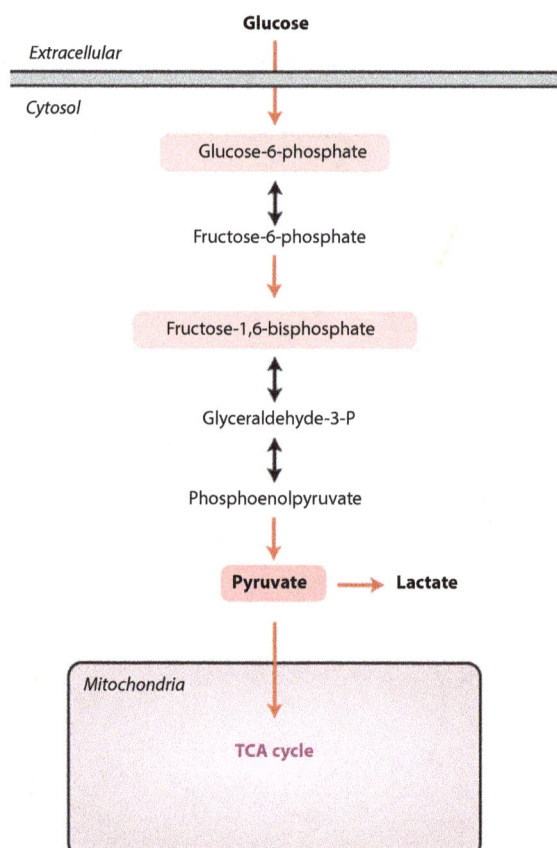

Figure 5-3. Breakdown of glucose.

Table 5-1. Glucose facilitators/transporters.

Protein	Insulin-regulated?	Tissues
GLUT1	No	Red blood cells, cornea, brain, placenta
GLUT2	No	*β-islet cells*, liver, kidney, small intestine. Imports monosaccharides from gut enterocytes into blood.
GLUT3	No	Brain, placenta
GLUT4	**Yes**	Adipose tissue, striated muscle
SGLT1	No	GI tract, renal reabsorption of glucose from proximal tubule
SGLT2	No	Renal reabsorption of glucose (proximal tubule)

4. Glucose entry into the cell

a. Glucose enters the cell through facilitators (**GLUT** or sometimes **SGLT** proteins), and the cell needs a way to trap it there so it can't exit the cell

   1. Different cell types use different facilitator proteins
   2. Step 1 can test this, since it relates particularly to insulin response and diabetes
   3. GLUT tranporters use facilitated diffusion, while SGLT is active co-transport using Na^+

b. Glucokinase and hexokinase tag the glucose with a phosphate to trap it in the cell and their regulation is a key detail to know

   1. Glucokinase: Found in cells sensitive to high levels of glucose: Liver and pancreatic β-cells
      - Low affinity, high capacity—sensitive to glucose levels and able to process large intake
   2. Hexokinase: Found in virtually all cell types
      - High affinity, low capacity—using glucose directly for energy; enables glucose uptake even during fasting at constant, relatively low levels

c. Other properties of these enzymes follow the same pattern

   1. Glucokinase activity is increased by insulin—transcription increases, inducing increased glucose uptake in insulin-sensitive cells in the fed state
   2. Hexokinase is inhibited by glucose-6-phosphate (product inhibition)— avoids taking in more glucose than necessary

5. Glycolysis

a. All cells in the body perform glycolysis, which produces, pyruvate, ATP, and NADH

b. Pyruvate is processed into different pathways depending on energy needs of the body

   1. **Aerobic catabolism:** In most cells, pyruvate is used in the TCA cycle to produce additional NADH under aerobic conditions
      - NADH is used to produce ATP through oxidative phosphorylation when oxygen is available
      - Oxidative phosphorylation requires mitochondria

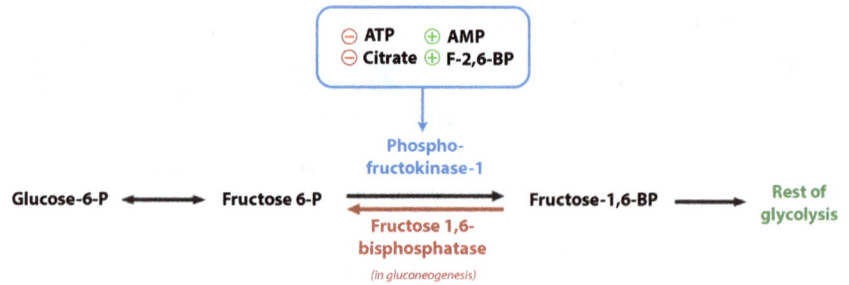

Figure 5-4. Phosphofructokinase-1 reaction in glycolysis.

      2. **Anaerobic catabolism:** If oxygen is not available, NADH is recycled by converting pyruvate to lactate
- Also used in cells without mitochondria (e.g., red blood cells)

      3. **Other processes:** Pyruvate can be used in gluconeogenesis to produce glucose or fatty acid synthesis (via acetyl-CoA)

    c. **Clinical:** Anaerobic metabolism predominants in a few examples particularly important for Step 1

      1. During exercise: Oxygen stores are depleted in intense exercise (e.g., sprinting)

      2. Red blood cells (erythrocytes): No mitochondria, so rely exclusively on glycolysis even in the presence of oxygen

      3. Cancer cells: Rapid growth rate — anaerobic glycolysis predominates

    d. Regulation of glycolysis: Fructose-6-phosphate (**F6P**) — fructose-1,6-bisphosphate (F-1,6-BP) is the rate-limiting step

      1. Phosphofructokinase-1 (PFK-1) catalyzes rate-limiting reaction
- Understand the principles that determine regulation here: Similar to regulation of TCA cycle and gluconeogenesis
- ATP inhibits the reaction—if energy is abundant, the cell does not want to be making more, but rather storing glucose for later use
- AMP activates the reaction—lack of energy should drive the cell to make more
- Citrate inhibits the reaction—think of citrate in the cytosol as a sign of "overflow" of the TCA cycle in mitochondria
- Fructose-2,6-bisphosphate (F-2,6-BP) activates the reaction—see below

      2. Phosphofructokinase-2 (PFK-2) regulates PFK-1
- Remember regulators by the "2" in their name (think primary vs. secondary pathway)

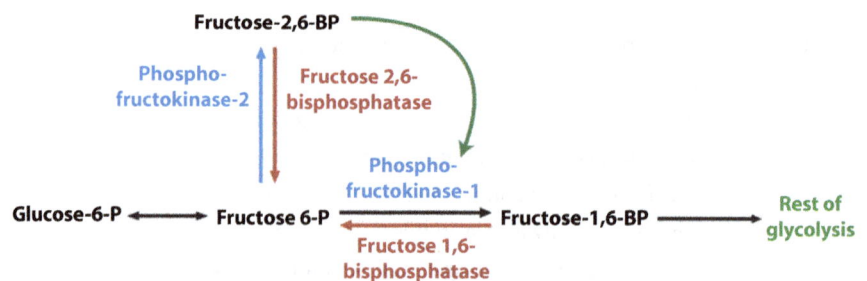

Figure 5-5. Regulation of glycolysis by phosphofructokinase-2.

- **PFK-2** catalyzes F6P → F-2,6-BP (not directly part of glycolysis pathway)
  a. F-2,6-BP activates PFK-1 → activates glycolysis
- Fed state—lots of glucose → plenty of F6P available to react with PFK-2
  a. Another unit of the enzyme, fructose 2,6-bisphosphatase (FBPase-2), converts F-2,6-BP back into F6P for normal processing
- Insulin signaling in liver induces PFK-2 (and enhances glycolysis)
  a. ↑ insulin → ↓ cAMP → ↓ protein Kinase A Activity
  b. In liver, phosphorylation by protein kinase A inhibits PFK-2 (and activates FBPase-2) → decreasing it activates PFK-2 and, hence, glycolysis
  c. This can seem like a lot of abbreviations to remember so ask: What would insulin have to do to activate glycolysis?

6. Pyruvate processing

a. Pyruvate dehydrogenase (PDH) — links glycolysis to TCA cycle and fatty acid synthesis
  1. This comes up less often than the PFK enzyme regulation, but it is the enzyme responsible for sending pyruvate to the TCA cycle, so it is also regulated
  2. Enzymes: PDH is a complex of multiple enzymes, but don't worry about what they are
  3. Cofactors: Energy production-related B vitamins (see Chapter 9)
     - Vitamin B_1 (thiamine)
     - Vitamin B_2 (FAD)
     - Vitamin B_3 (NAD)
     - Vitamin B_5 (CoA)
     - Lipoic acid
  4. Regulation: Think "high energy usage" — process immediately for maximum ATP production; activated by
     - ADP
     - NAD^+—used by TCA cycle to make NADH
     - Ca^{2+}—released in muscle cells during contraction
  5. **Clinical**
     - Arsenic inhibits lipoic acid and thus PDH, causing **drowsiness, confusion,** etc., problems with energy production hit the brain
     - Pyruvate dehydrogenase deficiency due to genetic mutation; most often X-linked; same neurologic effects
       a. Pyruvate is shunted to other pathways: **Lactic acidosis**, alanine buildup
       b. Treat with **ketogenic diet** to avoid pyruvate-related amino acids (see Chapter 9)

b. Lactate
  1. Glycolysis followed by the TCA cycle and oxidative phosphorylation produces much more ATP per glucose than glycolysis alone
  2. Glycolysis can still produce limited ATP on its own, but is limited by NAD^+ availability

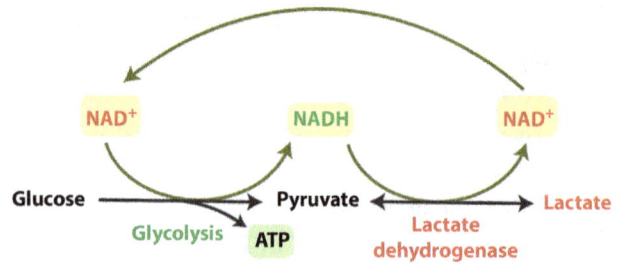

Figure 5-6. Lactate production.

3. Lactate dehydrogenase converts pyruvate into lactate and recycles NADH back to NAD⁺
4. Lactate can then be transported back to the liver, where it can be converted back to pyruvate and used in gluconeogenesis
5. **Clinical:** Lactic acidosis is an important cause of anion gap metabolic acidosis: Lactic acid leads to buildup of acid in the blood

7. **TCA cycle**

   a. The TCA cycle is the major source of ATP in cells using aerobic respiration
   b. Overall picture to take away: A cycle of carbon compounds allows **acetyl-CoA** to be processed through several steps
   c. **NADH** and **FADH₂** are produced as major sources of energy; GTP is also produced
   d. Regulation
      1. Several steps are regulated but you should not try to memorize them all
      2. Rather, think of what would regulate each one: Low energy states/low TCA usage
      3. ATP, NADH, and products inhibit reactions
      4. ADP activates reactions

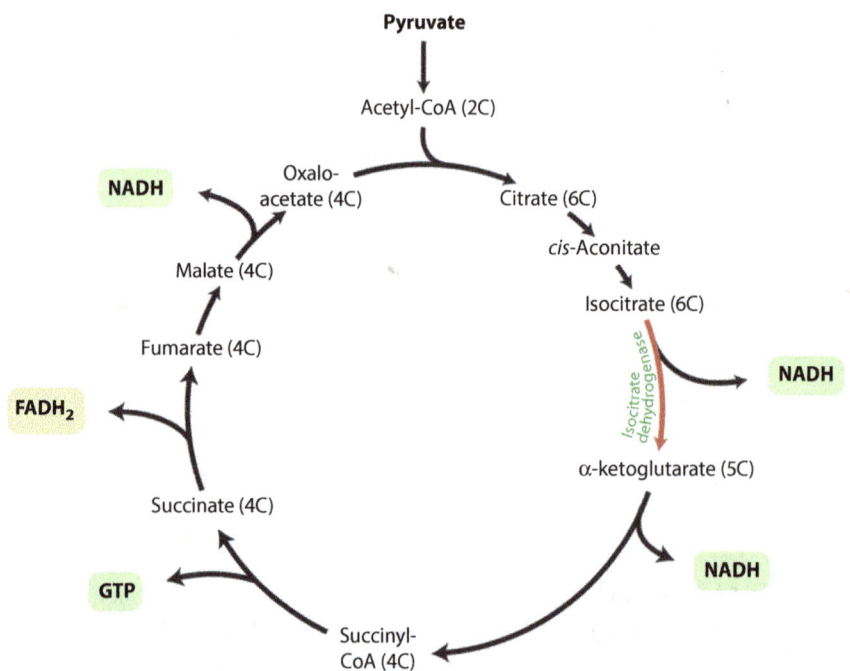

Figure 5-7. TCA cycle.

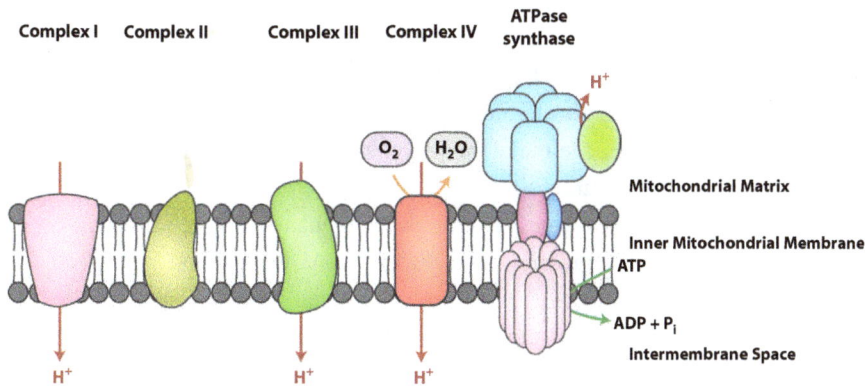

Figure 5-8. Oxidative phosphorylation.

  e. Going one step further
    1. **Isocitrate dehydrogenase** is the rate-limiting step
    2. **α-Ketoglutarate dehydrogenase** has the same cofactors as PDH
       (B_1, B_2, B_3, B_5, and lipoic acid)

8. Oxidative phosphorylation

  a. Oxidative phosphorylation involves conversion of NADH and $FADH_2$ into
     ATP in the mitochondria
  b. The key steps are
    1. Energy stored in NADH and $FADH_2$ is transferred through complexes
       in the inner mitochondrial membrane (the electron transport chain)
    2. With transfer of electrons, protons (H^+) are pumped into the very small
       intermembrane space, building up an electrochemical gradient
    3. Finally, H^+ are allowed out of the intermembrane space, but only if they
       power ATP synthase to produce ATP
  c. NADH (predominant product) uses Complexes I, III, and IV; while
     $FADH_2$ uses Complexes II, III, and IV
  d. Clinical: Oxidative phosphorylation is central to human life: Many serious
     poisons interrupt oxidative phosphorylation

Table 5-2. Inhibitors of oxidative phosphorylation.

Inhibitor	Target	Mechanism
Rotenone	Complex I	Inhibits reactions that build up H^+ gradient
Antimycin	Complex III	Inhibits reactions that build up H^+ gradient
CO, Cyanide	Complex IV	Inhibit reactions that build up H^+ gradient
Oligomycin	ATP synthase	Inhibits ATP synthase so no energy production, even though there is an H^+ gradient
2,4-dinitrophenol, aspirin, thermogenin	Decoupling agent	Makes membrane permeable, preventing buildup of H^+ gradient. Leads to heat buildup (fever, brown fat warmth)

9. Gluconeogenesis

a. Primarily occurs in the liver and kidney as a way to turn energy stores back into glucose for export to the rest of the body (in particular to the brain)

b. Gluconeogenesis is in many ways glycolysis in reverse, but there are 4 reactions that are unique

c. Having separate enzymes responsible for regulatory steps ensures that the body can regulate glycolysis and gluconeogenesis separately

d. Key enzymes: "Carboxy" or "phosphatase" related, as gluconeogenesis builds up carbons and removes phosphates to export glucose

   1. Pyruvate carboxylase—pyruvate converted to oxaloacetate (requires biotin, ATP)

   2. Phosphoenolpyruvate carboxykinase—oxaloacetate converted to phosphoenolpyruvate (requires GTP)

   3. **Fructose 1,6-bisphosphatase**—reverse reaction of PFK-1 from glycolysis and regulated by the same compounds

   4. **Glucose-6-phosphatase**—reverses hexokinase/glucokinase and is not present in all cells (e.g., muscle lacks it); required to export glucose

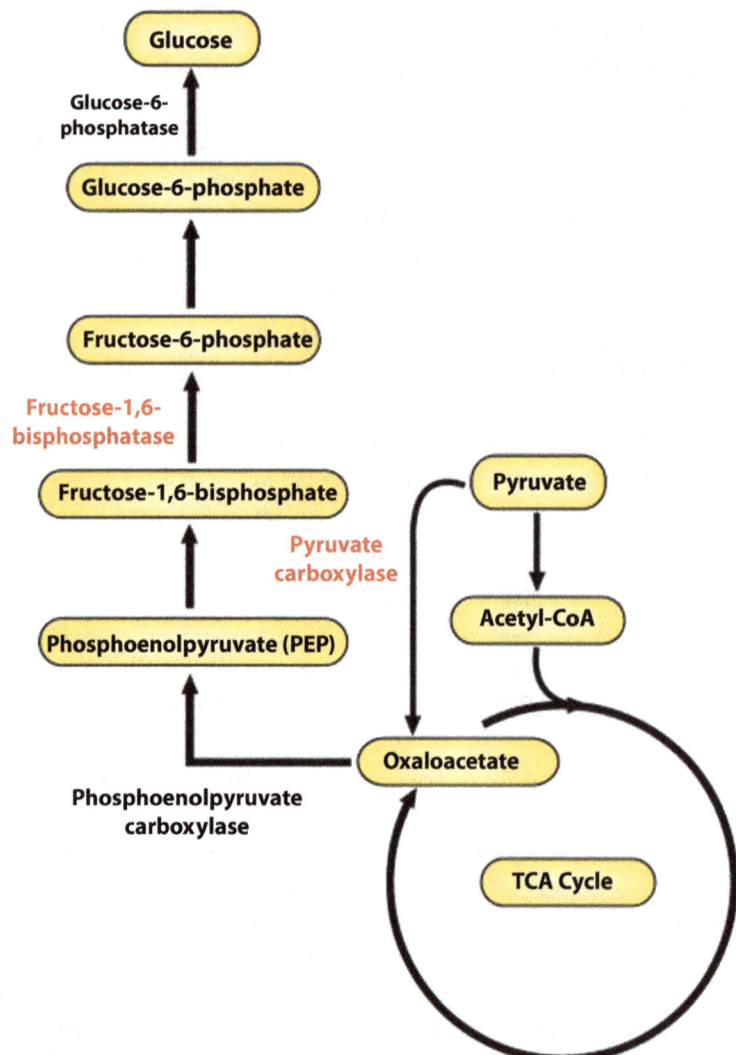

Figure 5-9. Gluconeogenesis.

e. **Clinical:** Deficiencies exist, though only one is commonly tested

1. **Von Gierke disease:** Deficiency of glucose-6-phosphatase
   - Severe fasting **hypoglycemia** → insulin levels fall
   - Liver cannot export glucose, so it is stored as glycogen and builds up → **hepatomegaly**
   - Buildup of glucose stores in the cell → increased utilization in glycolysis and pentose phosphate pathway results in **hypertriglyceridemia**, lactic acidosis, hyperuricemia
2. Other deficiencies rarely tested, but in general: Hypoglycemia
3. Due to regulation by insulin/glucagon, may also come up related to diabetes

10. Summary

a. Learning metabolism is as much about understanding where things fit together as about learning each pathway

b. **Regulation** of key steps is tested on Step 1 to ensure you understand the process, so there are worthwhile aspects to learn

c. This regulation is applied in the following chapters, which discuss carbohydrate (Chapter 6), amino acid (Chapter 7), and lipid (Chapter 8) metabolic pathways and diseases

Table 5-3. Summary of metabolic pathways.

Pathway	Location	Major diseases
Carbohydrates (Chapter 6)		
Non-glucose monosaccharides	All (cytosol)	Fructose intolerance Classic galactosemia
Glycogenesis/Lysis	Liver, muscle (cytosol)	Glycogen storage diseases (e.g., Von Gierke, Pompe)
Pentose phosphate	All (red blood cells clinically relevant)	Glucose-6-phosphate dehydrogenase (G6PD) deficiency
Amino acids (Chapter 7)		
Amino acid catabolism	All (cytosol → mitochondria)	Maple syrup urine disease Homocystinuria
Catecholamine synthesis	All (locally in brain)	Phenylketonuria
Urea cycle	Liver (mitochondria → cytosol)	Ornithine transcarbamylase deficiency
Lipids (Chapter 8)		
Lipogenesis/Lysis	Cytosol, mitochondria	Carnitine deficiency Medium-chain acyl-CoA dehydrogenase deficiency
Cholesterol synthesis	Liver	Hypercholesterolemia
Ketone bodies (Chapter 9)		
Ketone synthesis	Liver (mitochondria)	Diabetic ketoacidosis
Ketone catabolism	Muscle, brain	Diabetic ketoacidosis

    d. Although genetic disease is rare, many serious poisons affect the TCA cycle and oxidative phosphorylation
        1. **Arsenic** prevents initiation of the TCA cycle
        2. **Carbon monoxide** and **cyanide** both inhibit the oxidative phosphorylation and are commonly tested, e.g., in factory workers
        3. **Aspirin** overdose leads to fever and inhibition of oxidative phosphorylation due to uncoupling of ATP synthase
    e. Understanding the purpose of each pathway will minimize the need for rote memorization (e.g., activation/inhibition of rate-limiting enzymes)

Practice questions

1. A new mother presents to the emergency department with her 3-day-old son, stating that he refuses to eat and is always crying. Blood tests show high levels of fatty acids, glucose, and other nutrients in the blood, but the baby has lost weight since birth. Further analysis indicates impaired acetyl-CoA metabolism, and the physician suspects an environmental toxin is at fault. Why is buildup of acetyl-CoA a serious health risk?

    A. Acetyl-CoA forms intracellular aggregates, which are cytotoxic.
    B. Acetyl-CoA promotes amino acid breakdown and excess leads to dangerous muscle wasting.
    C. Buildup of acetyl-CoA signals arrest metabolic pathways used to produce energy.
    D. Without acetyl-CoA metabolism, all products of glycolysis are shuttled to lipogenesis and oxidative phosphorylation cannot occur.
    E. Without acetyl-CoA metabolism, energy cannot be stored via lipogenesis and glycogen synthesis.

2. A phenotypically normal couple has 3 children: 1 son and 2 daughters. All 3 children suffer from a disorder of mitochondrial DNA that leads to myopathies, encephalopathy, and a host of other symptoms. In these patients, metabolic functions of the mitochondria are significantly impaired. A new clinical trial uses gene therapy to treat the disorder by repairing the mutated gene. Levels of lactate, pyruvate, and acetyl-CoA are measured as outcome metrics. If the therapy is successful, what changes are expected in the cells of patients receiving treatment?

    A. Lactate decreases, pyruvate decreases, acetyl-CoA decreases
    B. Lactate decreases, pyruvate decreases, acetyl-CoA increases
    C. Lactate decreases, pyruvate increases, acetyl-CoA increases
    D. Lactate increases, pyruvate decreases, acetyl-CoA increases
    E. Lactate increases, pyruvate increases, acetyl-CoA decreases

3. A 71-year-old man is diagnosed with advanced pancreatic cancer, for which there is no approved treatment. He is given a 6-month prognosis. However, his oncologist informs him that he is a candidate for a study of a new chemotherapy drug. The drug inhibits lactate dehydrogenase, and has shown efficacy in animal models of reducing tumor sizes. The patient enrolls in the study. What

is the most likely mechanism by which this drug affects cancer survival and growth?

A. It blocks production of growth factors, inducing apoptosis
B. It inhibits anaerobic glycolysis
C. It inhibits beta-oxidation of lipids
D. It inhibits entry of non-glucose monosaccharides into the glycolysis pathway
E. It is an inhibitor of the citric acid cycle

4. A cancer biologist is studying the interaction between metabolism and apoptosis mechanisms across different cell types, in an effort to identify new drug targets to limit cancer cell growth. In one part of the study, she collects samples of different blood cell populations from cancer patients and controls. After purifying each cell type, she lyses the cells and examines their metabolic contents using high performance liquid chromatography (HPLC). If she were examining a population of red blood cells (erythrocytes), which of the following compounds would she expect to find most enriched, relative to the other compounds listed?

A. Citrate
B. Fatty acyl-carnitine
C. Acetyl-CoA
D. Pyruvate
E. Triglycerides

5. A cancer biologist is investigating new avenues for chemotherapy that focus on inhibiting cellular metabolism. She has identified glycolysis, a major pathway in anaerobic energy production, as a promising target. She develops an inhibitor of phosphofructokinase-1 (PFK-1), which catalyzes the rate-limiting reaction. When applied to cancer cells, this drug leads to significant ablation of metabolic activity. Drugs targeting other enzymes in the glycolysis pathway do not have a significant effect on cancer cell growth. Why does inhibiting PFK-1 provide the strongest result?

A. Each of the other steps of glycolysis is mediated by multiple enzymes that cannot be inhibited at once.
B. Only enzymes that mediate rate-limiting reactions contain regulatory sites to modulate reaction rates.
C. Production of rate-limiting enzymes is inhibited until induction by downstream feedback mechanisms.
D. The PFK-1 mediated rate-limiting reaction has a positive change in Gibbs free energy, so functional enzyme is necessary for the reaction to proceed.
E. The reaction catalyzed by PFK-1 proceeds at a slower rate than any other reaction, so it determines the rate of the overall process.

6. A 31-year-old factory worker asks her primary care physician for a sleeping aid because she consistently feels tired and dizzy at work. She works in a factory that utilizes cadmium in large quantities. The physician suspects this may be the problem, as cadmium can interfere with oxidative phosphorylation. In particular, it can lead to increased formation of reactive oxygen species (ROS), which

can cause many problems within the cell. What is the specific function of oxygen in the electron transport chain?

A. Electron transport drives the production of an oxygen gradient across the mitochondrial membrane, which drives ATP production.
B. It functions as an electron donor to Complex I.
C. It is a byproduct of the reaction, as the reaction derives protons from the splitting of water to create a proton gradient and produce ATP.
D. It is a co-factor necessary for function of Complex III.
E. It is the final electron acceptor of the electron transport chain.

7. Researchers at a military hospital are investigating potential chemical weapon threats. They hope to develop rapid-acting antidotes that can be carried by medics for rapid administration on the field. In particular, one researcher is studying the activity of cyanide, which can induce seizures and respiratory arrest within seconds. By what mechanism does cyanide exert its detrimental effects?

A. Activation of fructose 2,6-bisphosphatase
B. Inhibition of ATP synthase
C. Inhibition of hexokinase
D. Inhibition of pyruvate dehydrogenase
E. Inhibition of the electron transport chain

8. A 58-year-old man with type II diabetes regularly measures his blood glucose level to ensure that he is properly managing his disease. Even when he has not eaten in several hours, his blood glucose level remains at 130 mg/dL. Which of the following cellular sources of glucose could contribute to this elevated fasting blood glucose level?

A. Glucokinase in liver cells
B. Glucokinase in muscle cells
C. Glucose-6-phosphatase in muscle cells
D. Glucose-6-phosphatase in kidney cells
E. Glycogen phosphorylase in muscle cells

9. A 62-year-old man with a 1-year history of type II diabetes mellitus presents to his primary care clinic for a routine checkup. On metformin therapy and lifestyle counseling, his HbA_{1c} is 8.0% (goal ≤7%), which is unchanged since his initial diagnosis. He states that he is compliant with the medication. The physician recommends beginning a long-acting insulin in the morning. Increasing insulin availability in the body will lead to increased activity of which of the following enzymes?

A. Fructose-1,6-bisphosphatase
B. Glucose-6-phosphatase
C. Hexokinase
D. Phosphofructokinase-2
E. Pyruvate carboxylase

10. A medical research group studies diabetes mellitus. A scientist in the laboratory is investigating the different types of glucose transporters, hoping to find new drugs that could mimic the action of insulin to stimulate uptake of glucose by

muscle cells. Which of the following transporters would be a good candidate for such a treatment focused on treatment of diabetes mellitus?

A. GLUT-1
B. GLUT-2
C. GLUT-3
D. GLUT-4
E. SGLT-1

Carbohydrate Metabolism

CHAPTER 6

1. Purpose

a. Glycolysis and the TCA cycle are the major pathways of energy production in the body.

b. The body can't just rely on glucose—other carbohydrates, proteins, and lipids are also important sources of energy

c. The next 4 chapters will cover these additional pathways involved in energy production and storage

d. Diseases resulting from dysfunction of these side pathways are a big part of how the Step 1 likes to test metabolism

2. Strategy

a. Each side pathway has a function: What does the body need it for?

b. Symptoms often relate to limiting the body's ability to perform this function

c. Don't memorize every enzyme/step of the pathways—focus on the 1–2 enzymes per pathway that are tested

d. Use mnemonics for groups of diseases like glycogen storage, and make your own if you need to

3. Overview of metabolic pathways

a. Chapter 5 focused on the main path from glucose to ATP production (glycolysis, TCA cycle, and oxidative phosphorylation) as well as glucose synthesis by gluconeogenesis

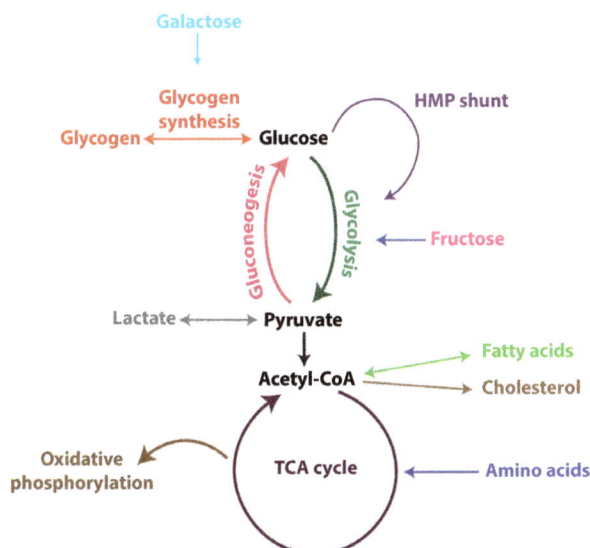

Figure 6-1. Overview of metabolic pathways.

Table 6-1. Summary of pathways.

Pathway	Location	Major diseases
Carbohydrates (Chapter 6)		
Gluconeogenesis	Liver (mitochondria → cytosol → ER)	Hypoglycemia
Non-glucose monosaccharides	All (cytosol)	Fructose intolerance Classic galactosemia
Glycogenesis/Lysis	Liver, muscle (cytosol)	Glycogen storage diseases (e.g., Von Gierke, Pompe)
Pentose phosphate	All (red blood cells clinically relevant)	Glucose-6-phosphate dehydrogenase (G6PD) deficiency
Amino acids (Chapter 7)		
Amino acid catabolism	All (cytosol → mitochondria)	Maple syrup urine disease Homocystinuria
Catecholamine synthesis	All (locally in brain)	Phenylketonuria
Urea cycle	Liver (mitochondria → cytosol)	Ornithine transcarbamylase deficiency
Lipids (Chapter 8)		
Lipogenesis/Lysis	Cytosol, mitochondria	Carnitine deficiency Medium-chain acyl-CoA dehydrogenase deficiency
Cholesterol synthesis	Liver	Hypercholesterolemia
Ketone bodies (Chapter 9)		
Ketone synthesis	Liver (mitochondria)	Diabetic ketoacidosis
Ketone catabolism	Muscle, brain	Diabetic ketoacidosis

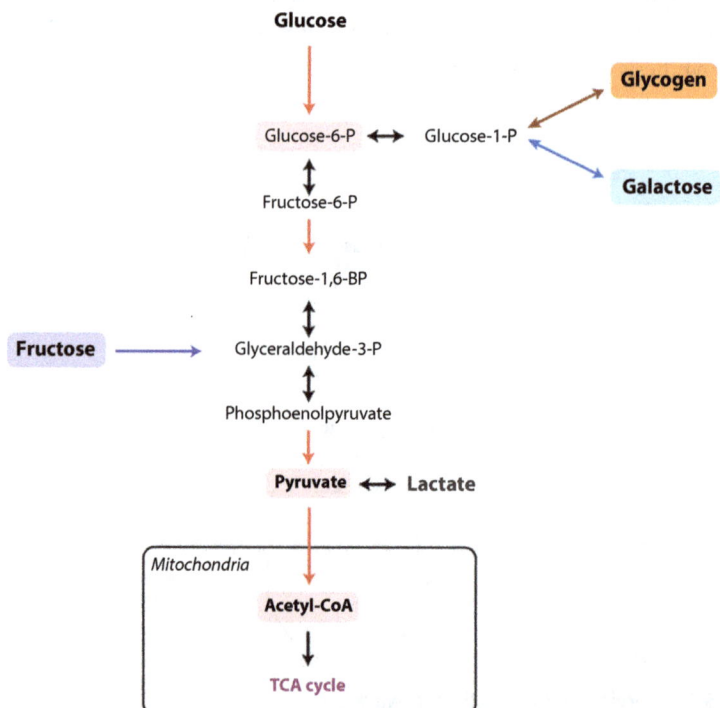

Figure 6-2. Links between glucose metabolism and other carbohydrates.

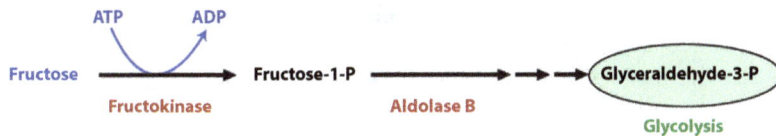

Figure 6-3. Fructose metabolism.

    b. Chapter 6 focuses on other carbohydrates
       1. **Fructose** and **galactose** are common dietary carbohydrates that enter into the glycolysis pathway for energy production
       2. **Glycogen** is a polysaccharide, consisting of multiple glucose subunits, for energy storage

4. Fructose metabolism

    a. Fructose is found in "naturally sweet" things—e.g., fruit, honey
    b. Both fructose and galactose (next section) follow very similar process that you can think of in 2 main steps
       1. First, like glucose they have to be phosphorylated and trapped in the cell: Fructokinase does this
       2. Then, it is converted by aldolase B (and others) through intermediates into a compound used in glycolysis (glyceraldehyde-3-P)
    c. The actual process has more steps in it, but it's highly likely Step 1 will focus on the steps where deficiency leads to specific diseases
    d. **Clinical:** Two diseases to know, each associated with a single step
    e. Deficiency of **fructokinase** leads to essential fructosuria
       1. Fructose shows up in the urine—the name is all you need
       2. Fructokinase traps fructose in the cell; without it fructose can get back out and will go unused into the urine
       3. No major symptoms except elevated fructose in blood and urine
    f. Deficiency of aldolase B leads to fructose intolerance
       1. As the name suggests, it is more serious
       2. Lack of aldolase B leads to accumulation of fructose-1-P in the cell and depletion of phosphate which is needed for other processes
       3. Symptoms: **Hypoglycemia**, **liver damage**, vomiting
    g. **Treatment:** No fructose or sucrose!

5. Galactose metabolism

    a. Galactose is present in milk (lactose)
    b. The first step of galactose metabolism is very similar to that of fructose
       1. Just like glucose and fructose, galactose is trapped in the cell by galactokinase to galactose-1-P
       2. It is then converted into glucose by uridyltransferase by swapping out galactose for glucose from the UDP cofactor
       3. In other words, 2 main steps in each reaction achieve the same core functions

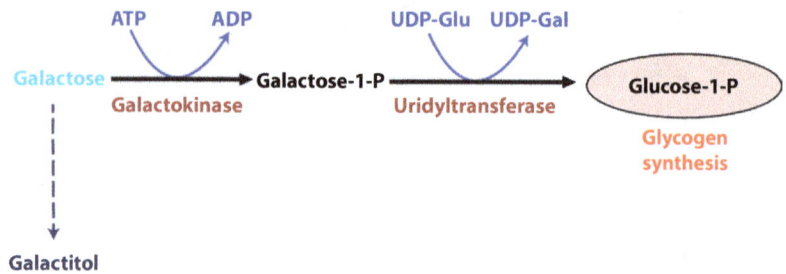

Figure 6-4. Galactose metabolism.

   c. These are the 2 steps that relate to disease; like in the case of fructose, there are other steps that are rarely tested

   d. Clinical: In general, they parallel fructose disorders but are more serious

   e. Deficiency of galactokinase leads to galactokinase deficiency

     1. Unlike fructose, galactose can be processed in a secondary reaction into **galactitol**

     2. Mostly benign as most galactose ends up in the urine without causing damage

     3. In infants who are more susceptible and slower to process toxins, buildup of galactitol can lead to **cataracts**

   f. Deficiency of uridyltransferase is more serious with buildup of multiple toxins causing classic galactosemia

     1. General **failure to thrive,** vomiting

     2. **Jaundice** and hepatomegaly

     3. Infantile **cataracts**

     4. Intellectual disability

     5. Risk of *E. coli* sepsis ("-emia" one way to think of this is we know bacteria like lactose, and lactose contains galactose)

   g. Treatment: No lactose or galactose!

6. Sorbitol

   a. As with galactose, sorbitol exists as a secondary way to process glucose to trap it in the cell, but can cause problems if it builds up

   b. Glucose is converted into sorbitol by aldose reductase, and then normally to fructose by sorbitol dehydrogenase

   c. Some tissues (retina, **lens,** Schwann cells, kidneys) can't process sorbitol

     1. Buildup causes osmotic damage (swelling)

   d. **Clinical: Cataracts,** retinopathy, peripheral neuropathy

   e. Seen in **diabetics** because of abnormally high blood glucose levels

Figure 6-5. Sorbitol.

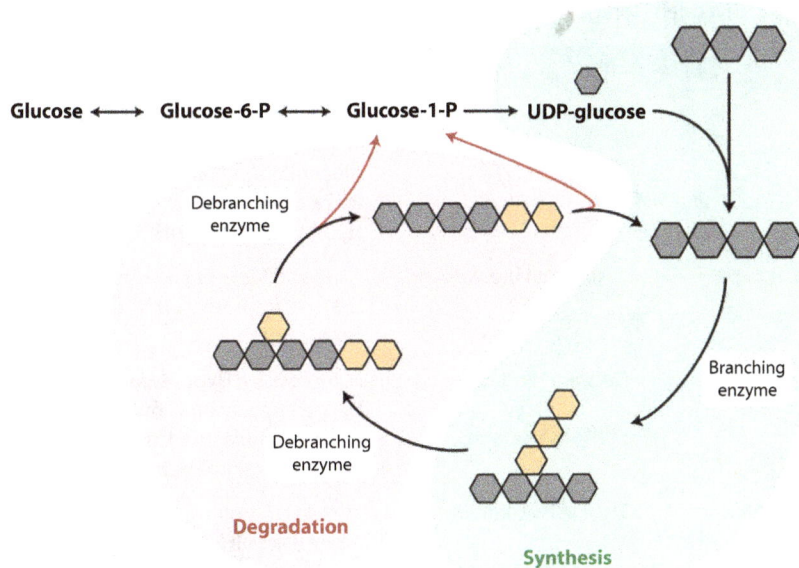

Figure 6-6. Glycogen synthesis and breakdown.

7. Glycogenesis

    a. Let's move into the energy storage pathways; the main one to learn is glycogen synthesis, which is polymerized glucose

    b. Crucial for

        1. **Muscle** cells—need a large energy reserve easily accessible

        2. **Liver** cells—break down glycogen to provide glucose to the rest of the body (in particular to the brain) during fasting

    c. Key synthesis steps

        1. Lots of enzymes involved; not high yield to learn each one; instead understand the overall process

        2. First, prepare glucose for polymerization by attaching UDP

        3. α-(1,4) bonds link one glucose to the next along opposite ends of the molecule

        4. For space efficiency, the strands can branch via α-(1,6) bonds (adjacent carbons)

    d. Key breakdown steps

        1. Remove a "standard" α-(1,4) linkage with glycogen phosphorylase

        2. Remove branches through 2-step process, both with debranching enzymes

            • Down to a manageable number of residues by transferring to the main chain

            • Specifically cleave α-(1,6) linkage

        3. Lysosome can directly break down the full chain into glucose molecules

    e. **Clinical:** There are several genetic diseases in the pathway that Step 1 likes to test

        1. Severity generally worse for lower numbers and more downstream (fewer "workarounds")

        2. **Diagnosis: Periodic acid-Schiff** (detects polysaccharides)

Table 6-2. Glycogen storage diseases.

	Disease	Step affected	Symptoms
I	Von Gierke	Glucose production	Fasting hypoglycemia, lactic acidosis Hepatomegaly (impaired gluconeogenesis) Elevated cholesterol, triglycerides Hyperuricemia (gout)
II	Pompe	Lysosomal breakdown	Cardiac effects (myopathy) Exercise intolerance, hypotonia Early death
III	Cori	Debranching	Milder type I: Hypoglycemia Hepatomegaly (improves with age) Elevated cholesterol, triglycerides Lactic acidosis/hyperuricemia **rare**
V	McArdle	Muscle breakdown	Muscle cramps, myoglobinuria with intense activity "Second wind" (↑ blood flow, fatty acids) Glycogen accumulation

3. Liver and muscle often affected (liver: more severe as more widespread consequences)
4. **Treatment:** Usually **regular glucose intake** to prevent the need to utilize glycogen in the first place

8. Pentose phosphate pathway

a. The pentose phosphate pathway (also called HMP shunt) uses glucose to produce 2 things
   1. **Ribose**, a carbohydrate of 5 carbons (pentose) for nucleotide synthesis
   2. **NADPH**, a cofactor related to NADH (vitamin B$_3$)
b. It takes place in the cytoplasm (like glycolysis)
c. Functions: Three main ones → only one is clinically relevant for Step 1
   1. Nucleotide synthesis: Think "pentose phosphate" → ribose
   2. NADPH production: Reducing agent → scavenges free radicals
   3. Lipid synthesis: Requires NADPH → fatty acids, cholesterol, and steroids
d. Key pathway steps
   1. Oxidative phase: Produces NADPH
      - Irreversible
      - **Glucose-6-P dehydrogenase** is the rate-limiting step
      - Feedback inhibited by NADPH

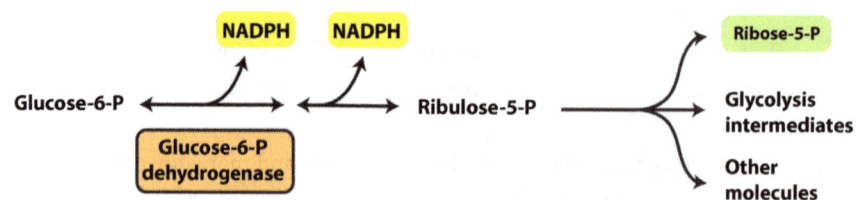

Figure 6-7. Pentose phosphate pathway (HMP shunt).

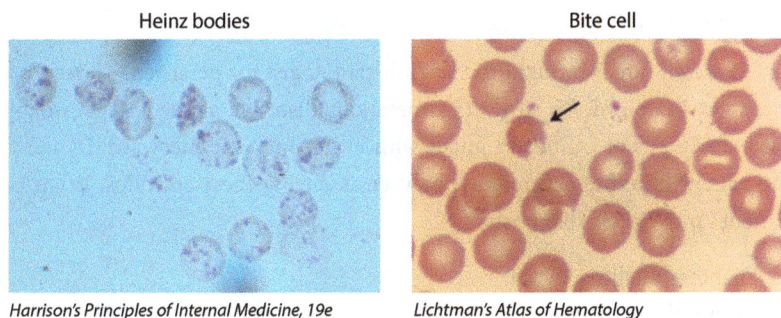

Heinz bodies

Bite cell

Harrison's Principles of Internal Medicine, 19e

Lichtman's Atlas of Hematology

Figure 6-8. Characteristic blood smears in G6PD deficiency.

   2. Nonoxidative phase: Produces ribose and other carbohydrates
- Reversible
- **Vitamin B$_1$** required as cofactor

e. **Clinical:** Pentose phosphate pathway is important in red blood cell function — only pathway to produce NADPH in red blood cells

   1. **Glucose-6-phosphate dehydrogenase deficiency** is the most common enzyme deficiency
- X-linked recessive
- Incidence correlated with regions where malaria has been endemic — trait hypothesized to confer **malaria resistance**

   2. Mechanism: RBCs need to prevent free radical buildup (glutathione)

   3. Causes: Patient has either started a new diet/drug, or gotten sick otherwise
- **Fava beans**, **sulfa drugs** (e.g., trimethoprim/sulfamethoxazole), primaquine (anti-malarial)
- **Infection** increases circulating free radicals

   4. Presentation: Related to lysis of RBCs
- **Anemia** (weakness, malaise)
- **Jaundice** (dark urine)
- Blood smear: **Heinz bodies**, **bite cells** — oxidative damage denatures hemoglobin — damaged RBCs are chewed up by macrophages

   5. **Treatment:** Avoid exposures; supportive treatment if necessary

9. Disease relevance summary

a. Carbohydrate diseases can be severe when compounds accumulate to high levels and/or in sensitive cells (**glycogen storage**, **sorbitol**)

b. Symptoms usually relate to energy deficiency

c. In childhood, disease can also have neurologic symptoms (in developing brain especially)

d. Diseases within each class are easy to group together (**Von Gierke** vs. **Cori**)

e. Minimal pharmacology here: If there is a deficiency, it is corrected by dietary intervention

Practice questions

1. A 2-year-old boy is brought to the endocrinologist for persistent dietary issues. The referring physician suspects a metabolic disorder, as the child is unable to eat many types of fruits and juices without feeling sick afterward. In addition, the patient becomes severely hypoglycemic during these episodes. What is the patient's most likely diagnosis?

   A. Diabetes mellitus, type I
   B. Essential fructosuria
   C. Essential pentosuria
   D. Hereditary fructose intolerance
   E. Hereditary galactosemia

2. A 28-year-old gravida 1, para 1 woman gave birth to a son via Caesarian section and is recovering in the hospital. On the second day, after breastfeeding, her son becomes distressed and begins vomiting. The physician is concerned about a genetic disorder. Which of the following enzymes is most likely deficient in this patient?

   A. Aldose reductase
   B. Debranching enzyme
   C. Fructokinase
   D. Glucose-6-phosphate dehydrogenase
   E. Galactose 1-phosphate uridyltransferase

3. A 71-year-old man presents to the diabetes clinic with vision problems. The clinician on staff refers him to ophthalmology, and a cataract surgery is performed. The ophthalmologist informs the patient that diabetes was directly linked to his cataract formation. What is the molecule most likely responsible for this condition?

   A. Advanced glycation end-products
   B. Ethanol
   C. Fructose
   D. Sorbitol
   E. Insulin

4. A cardiologist who runs a basic science research lab is studying energy production by the heart. He hopes to identify factors that are protective against ischemia. To approach this question, he treats exercising rats with small molecules, one at a time, each of which is predicted to inhibit a different enzyme involved in energy metabolism. He then collects a sample of cardiomyocytes and examines their energy stores. After treatment with a particular small molecule, he notes changes in glycogen content. Specifically, the mass spectrometric analysis indicates that there is a relative increase in the number of α-(1,6) glycosidic bonds within glycogen. What enzyme is this molecule most likely inhibiting?

   A. Branching enzyme
   B. Debranching enzyme
   C. Glucose-6-phosphatase
   D. Glycogen phosphorylase
   E. Glycogen synthase

5. A 12-year-old girl is brought to the clinic complaining of weakness and fatigue. She wanted to join a junior soccer team, but was unable to complete the tryouts. She says that it was cold outside that day, and her muscles kept cramping, which made it difficult to play. Since then, she's been going on jogs with her mother to try to increase her endurance. The physician orders labs, including a urine test that identifies the presence of myoglobin. She suspects an enzyme deficiency is causing this patient's symptoms. Which of the following is the most likely diagnosis?

   A. Cori disease
   B. Galactokinase deficiency
   C. McArdle disease
   D. Pompe disease
   E. Von Gierke disease

6. A 5-month-old boy is brought in for evaluation by his parents due to muscle weakness. He has not been reaching milestones for motor function. The physician makes a diagnosis and explains that the patient suffers from a lysosomal defect in glucose production. His prognosis is grim, as he will most likely die from infantile heart failure within the next 2 years. What is the most likely diagnosis?

   A. Cori disease
   B. Gaucher disease
   C. Pompe disease
   D. Tay-Sachs disease
   E. Von Gierke disease

7. A 2-month-old girl is brought to the pediatrician's office because the parents are concerned about her motor development. She cannot support her head, and in general seems very weak. Blood tests show hypoglycemia and hyperlipidemia, lactic acidosis, and hepatomegaly. Further analysis shows that her cells lack glucose-6-phosphatase, meaning that she cannot derive glucose from crucial intracellular energy stores. What is the most likely diagnosis?

   A. Cori disease
   B. McArdle disease
   C. Pompe disease
   D. Tay-Sachs disease
   E. Von Gierke disease

8. A 34-year-old male patient comes to the clinic complaining of a new skin rash that appeared on his upper thigh. He worries that he got it from using uncleaned gym equipment, as he has been on a new workout regimen that includes weightlifting. The physician examines the rash and takes a skin sample and sends it for evaluation. Staining shows gram-positive cocci, and follow-up susceptibility testing indicates methicillin resistance. The physician considers antibiotic treatment options, but notes from the patient's chart that he has glucose-6-phosphate dehydrogenase deficiency. Which of the following drugs should be avoided in this patient?

   A. Clindamycin
   B. Doxycycline
   C. Linezolid
   D. Sulfamethoxazole
   E. Vancomycin

9. A pharmaceutical company whose major product is about to lose its patent is considering entering new areas of drug development. Its scientific director recommends producing food additives, as the market for these products is high and they are relatively cheap to develop. For example, antioxidants are often added to food as preservatives. What is the function of antioxidants in this context?

A. Buffer pH to prevent acid buildup
B. Kill bacteria by inhibition of oxidative phosphorylation
C. Inhibit bacterial growth (bacteriostatic)
D. Inhibit proteases that break down natural food products
E. Scavenge free radicals to prevent reactive oxygen species formation

10. Malaria, a disease caused by Plasmodium protozoa, affects red blood cells. Plasmodia replicate within RBCs, and eventually cause hemolysis and release of protozoa for infection of other RBCs. Glucose-6-phosphate dehydrogenase deficiency has been shown to protect individuals against malaria infection, providing selective pressure that has made this one of the most common enzyme deficiency in humans. What is the most likely explanation for how this deficiency protects against malaria?

A. Activation of apoptosis pathways
B. Decreased energy production for use by Plasmodium organisms
C. Decreased extracellular signal proteins needed for Plasmodium entry
D. Increased bicarbonate production leads to decreased pH
E. Increased erythrocyte sensitivity to free radicals disrupts Plasmodium life cycle

Amino Acid Metabolism

CHAPTER 7

1. Purpose

a. The third part of our metabolism review focuses primarily on amino acids and proteins

b. Amino acids are used in many pathways beyond protein synthesis, including energy production and neurotransmitter synthesis

c. Diseases in specific pathways do show up on Step 1, but overall, questions focus more on the function of each pathway and effects of diet on their function

2. Amino acid properties

a. Unlike in biochemistry classes, knowing specifics about each amino acid is not necessary for Step 1

b. What you learn depends on how much time you want to put into it

c. Charged amino acids: Highest value

   1. Basic (+): Lysine, histidine, arginine (*lies, hiss, arg*)
   2. Acidic (−): Glutamic acid (glutamate), aspartic acid (aspartate)

d. Polarity: Can help judge severity of genetic mutation

   1. Nonpolar: Glycine, alanine, valine, leucine, isoleucine, methionine, proline, phenylalanine, tryptophan
   2. Polar: All others (including acidic and basic)

e. Essential: Required in the diet vs. synthesized in the body

   1. Valine, leucine, isoleucine, phenylalanine, tryptophan, lysine, arginine, histidine, methionine, threonine
      - Grouping by similarity may help: Leucine/isoleucine; **phen**ylalanine, trypto**phan**; me**thio**nine, **threo**nine; histidine (basic), lysine (basic), and valine
      - Becomes important for common diseases, particularly phenylketonuria

3. Amino acids as energy

a. All amino acids can be used for energy if necessary

b. Amino acids can be glucogenic, ketogenic, or both, based on where they enter the pathway

   1. Glucogenic: Converted into compounds that can be used to make glucose by gluconeogenesis
      - Converted into pyruvate, or TCA cycle intermediates
   2. Ketogenic: Converted directly to acetyl-CoA or acetoacetate
      - Remember that neither acetyl-CoA nor ketone bodies can be used to make glucose
      - Leucine and lysine—exclusively ketogenic

Table 7-1. Glucogenic and ketogenic amino acids.

Glucogenic	Both	Ketogenic
Essential		
Histidine	Isoleucine	Leucine
Methionine	Phenylalanine	Lysine
Threonine	Tryptophan	
Valine		
Nonessential		
Alanine	Tyrosine	
Arginine		
Aspartate		
Asparagine		
Cysteine		
Glutamine		
Glutamate		
Glycine		
Proline		
Serine		

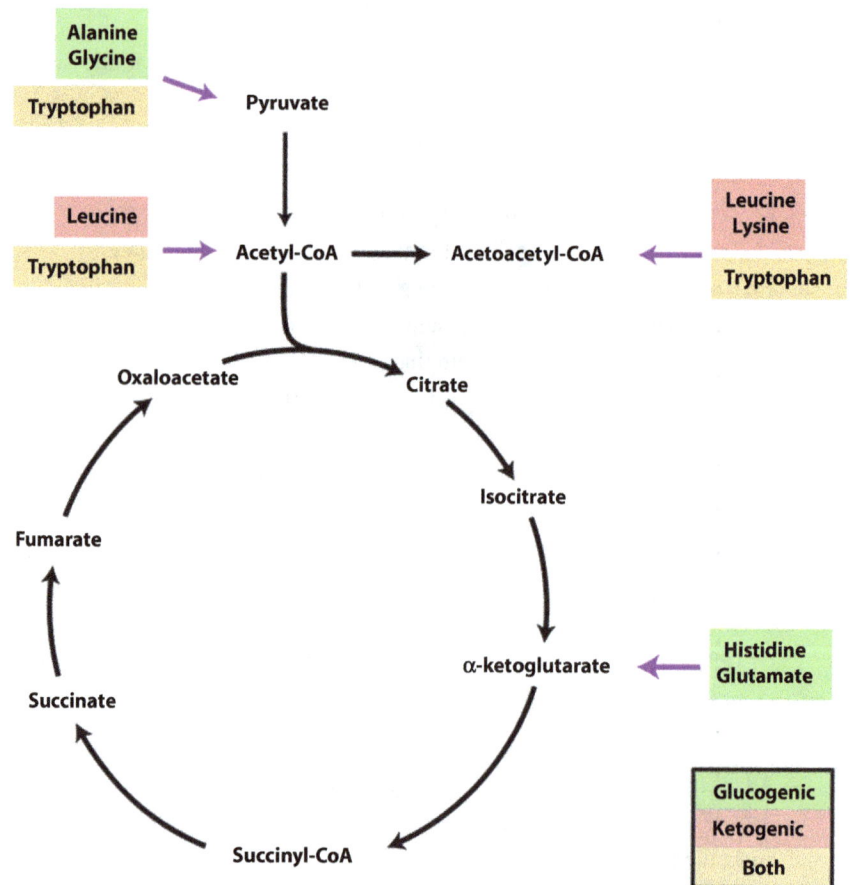

Figure 7-1. Examples of glucogenic and ketogenic amino acids.

c. **Clinical:** In **pyruvate dehydrogenase deficiency**, pyruvate cannot be converted to acetyl-CoA

1. Shunt to lactic acid leads to **acidosis** with neurologic symptoms
2. Appears in infancy: **Failure to thrive**, developmental delay, seizures
3. Treatment: Reduce glucogenic amino acids in diet, favor leucine and lysine

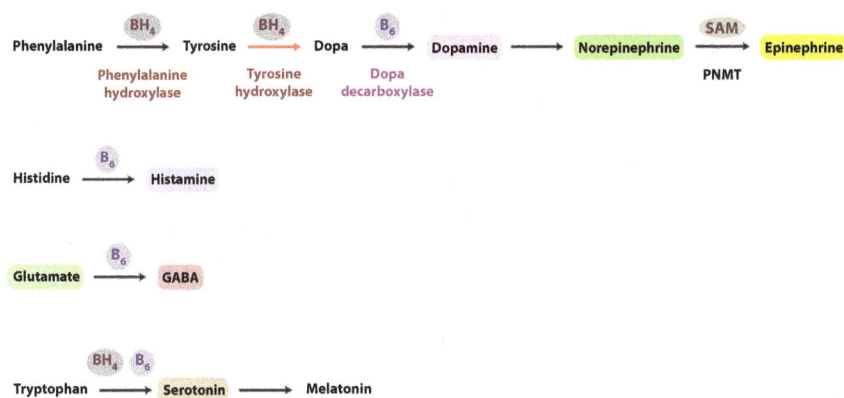

Figure 7-2. Neurotransmitter synthesis.

4.   Other amino acid metabolic pathways

a. We've reviewed amino acids' role in (1) protein synthesis (Chapter 3) and (2) energy production

b. There are several additional uses of amino acids, especially for neurotransmitter production

1. Knowing exactly which amino acids are made into which compounds is not important here

2. Some easy ones to remember are **histamine** and **glutamate/GABA**

3. These will come up in other organ systems when relevant to disease (e.g., heme production in porphyrias)

c. One key pathway that is tested is catecholamine synthesis from phenylalanine

1. **Phenylalanine** is an essential amino acid; converted to tyrosine

2. Tyrosine converted to L-dopa, then 3 neurotransmitters: **Dopamine**, **norepinephrine**, and **epinephrine**

   - Remember: Epinephrine is for acute situations, converted from NE when necessary

   - Norepinephrine (NE) → epinephrine (Epi) requires **pheylethanolamine *N*-methyltransferase (PNMT)**

3. Other pathways branch off

   - Tyrosine → melanin via **tyrosinase**

   - Tyrosine → homogentisic acid (break down)

   - Breakdown of catecholamines → by **catecholamine-*O*-methyltransferase (COMT)**

d. A few cofactors are used in these reactions

1. Tetrahydrobiopterin (BH_4) for early pathways that require hydroxylation

2. Vitamin B_6 for transamination reactions (histidine to histamine; dopa to dopamine)

3. S-adenosylmethionine (SAM) carries methyl groups (NE → Epi)

e. **Clinical:** Two diseases due to enzyme deficiencies are phenylketonuria and alkaptonuria (reviewed below)

2. Parkinson's disease results from decreased dopamine → treat to replenish dopamine

   - Dopamine can't cross the blood brain barrier → give **dopa** (also called L-dopa or levodopa), which will enter brain

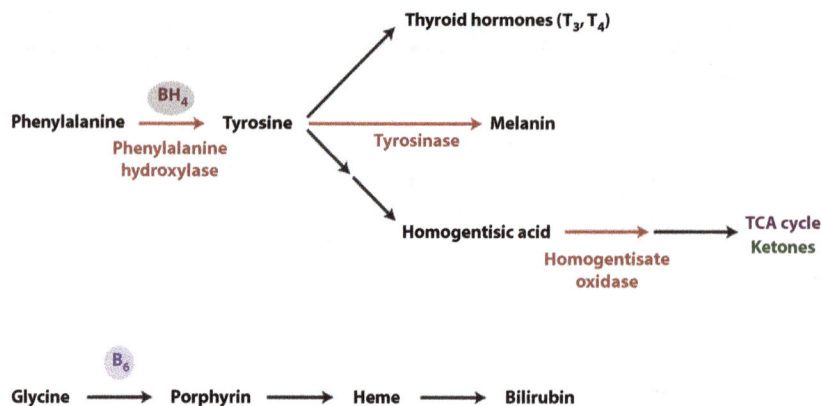

Figure 7-3. Other amino acid derivatives.

- Give **carbidopa** to inhibit peripheral dopa decarboxylase → reduce peripheral side effects of dopamine (nausea, vomiting, orthostatic hypotension)
3. **Albinism** typically results from mutation in tyrosinase → no melanin production
   - High risk of sunburn, **skin cancer**
   - Low visual acuity, photophobia

5. **Urea cycle**

   a. Use of amino acids for energy produces ammonia, which must be excreted → converted to urea to be removed by kidneys
   b. Takes place in the liver, in mitochondria (initial steps), and cytosol

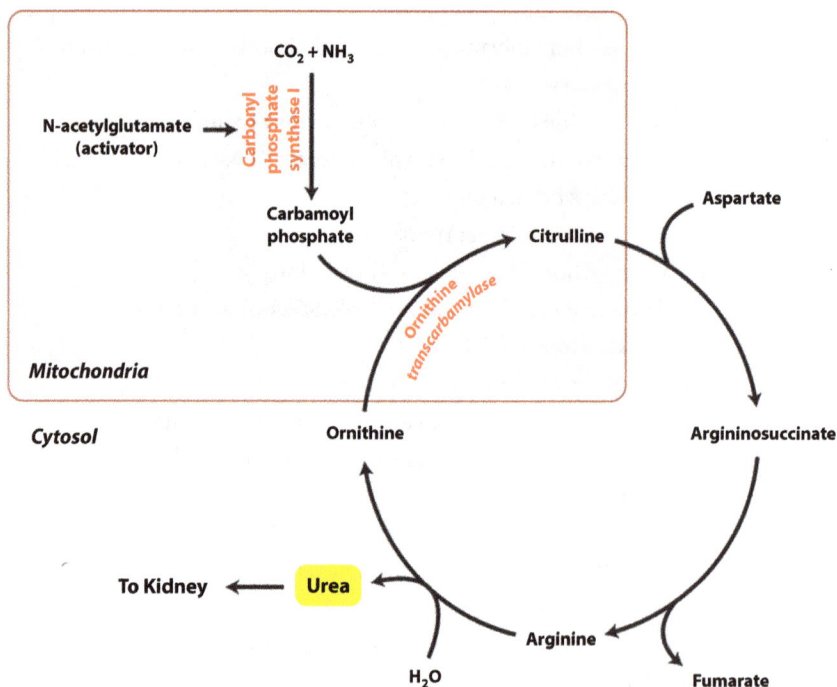

Figure 7-4. Urea cycle.

   c. It is not high yield for Step 1 to know every enzyme in the cycle: There are 1–2 enzymes to know

      1. **Carbamoyl phosphate synthase I**: Rate-limiting step, combines ammonia with bicarbonate ion and ATP

      2. **Ornithine transcarbamylase (OTC)**: Next step, most common deficiency

      3. Then, a cycle of compounds that leads to release of urea for export to kidneys

   d. **Clinical:** OTC deficiency is the disease Step 1 commonly uses to test the urea cycle

      1. X-linked recessive, shows up early in life

      2. Causes: Protein catabolism (part of normal metabolism)

      3. Mechanism: Impaired urea cycle leads to (1) increased ammonia and (2) increased carbamoyl phosphate which leads to pyrimidine synthesis and orotic acid buildup

      4. Presentation: **Hyperammonemia** is most severe
- Neurotoxin: Vomiting, confusion
- Tachypnea due to cerebral edema
- **Asterixis**: "Flapping" tremor when wrist is extended
- Also see increased urinary orotic acid
- Other causes of hyperammonemia: Liver failure → hepatic encephalopathy

      5. **Treatment:** Key is limit pathway by regulating protein intake, and keeping glucose intake high so body doesn't break down protein
- Episodes triggered by illness, fasting, etc.: **Phenylacetate** conjugates amino acids, excreted in urine
- Hemodialysis if necessary

6. Disorders of amino acid metabolism

   a. **Phenylketonuria**

      1. One of the most commonly tested disorders covered in this chapter
- Autosomal recessive
- All newborns are screened at 2 to 3 days old—treatable, and common among inborn errors of metabolism

      2. Causes: Mutations in **phenylalanine hydroxylase** prevent conversion of phenylalanine to tyrosine

      3. Mechanism: Phenylalanine can't be used → accumulation of phenyl ketones, interfering with brain development (exact mechanism unknown)

      4. Presentation: First years of life, often immigrant families (not automatically tested at birth)
- **Developmental delay** (intellectual disability)
- **Seizures**
- **Mousey odor** (ketones) in urine
- Eczema

      5. **Maternal PKU:** Mother with PKU → elevated phenylalanine levels early in pregnancy are exacerbated in the fetus
- Developmental delay, growth retardation, cardiac malformations
- Regardless of whether fetus has PKU
- Can appear similar to **fetal alcohol syndrome**

6. **Tetrahydrobiopterin (BH$_4$) deficiency:** Required cofactor for phenylalanine $\rightarrow$ tyrosine
   - Deficiency can appear similar to PKU
   - Additional effects: Related to impaired neurotransmitter synthesis
   - Motor deficits, progressive deterioration $\rightarrow$ death in infancy without rapid treatment
7. **Treatment:** Diet control $\rightarrow$ restrict phenylalanine, supplement tyrosine

b. Alkaptonuria

1. Causes: Homogentisate oxidase degrades tyrosine for energy production
2. Mechanism: Buildup of homogentisic acid $\rightarrow$ blueish pigmentation of skin, sclerae, etc.
3. Presentation: Usually benign
   - Urine turns black after exposure to air (black diapers)
   - Intervertebral calcium deposits seen on radiography
   - Arthritis due to deposits in cartilage
4. **Treatment:** Limit tyrosine and phenylalanine in diet
   - Limited efficacy: Can prevent progression but does not reverse existing damage

c. Maple syrup urine disease (MSUD)

1. Less often tested than PKU, but similarities in presentation
   - Autosomal recessive
   - $10\times$ rarer than PKU
2. Causes: Mutation in *branched-chain α-ketoacid dehydrogenase*
   - Branched-chain amino acids are isoleucine, leucine, and valine
3. Mechanism: Build up in the blood
4. Presentation: First years of life
   - **Developmental delay** (intellectual disability)
   - **Feeding problems,** vomiting
   - **Maple syrup** smell of urine
5. **Treatment:** Diet control; thiamine supplementation can sometimes boost enzyme activity

d. Homocystinuria

1. A collection of disorders all related to homocysteine metabolism
   - Also called **hyperhomocysteinemia**
2. Causes: Defects in enzymes processing homocysteine
   - Identifying root cause that determines correct treatment
3. Mechanism: Buildup of homocysteine in the blood and tissues
   - Homocysteine has pro-inflammatory and connective tissue effects
   - Effect is more defined than PKU/MSUD
4. Presentation: As early as 3 to 5 years old
   - Vascular/Inflammation: Prothrombotic $\rightarrow$ **atherosclerosis,** can lead to myocardial infarction, stroke
   - Connective tissue: **Lens subluxation**; Marfanoid habitus
   - Bone: **Osteoporosis**
   - Other: Developmental delay
5. **Treatment:** Depending on the enzyme, replace lacking substance
   - Methionine synthase: **Give methionine** in diet
   - Cystathionine synthase: **Give cysteine,** limit methionine
   - Decreased enzyme affinity for B$_6$: **Give B$_6$ and cysteine**

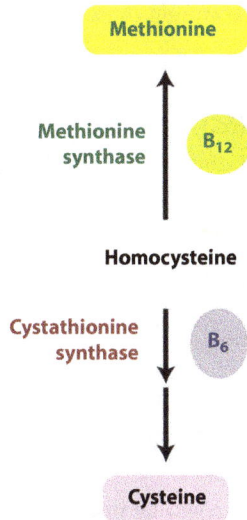

Figure 7-5. Homocysteine metabolism.

e. **Cystinuria**
   1. Often grouped with nephrolithiasis disorders (**kidney stones**)
      - Autosomal recessive
      - Other examples: Calcium oxalate, uric acid
   2. Causes: Defect of dibasic amino acid transporter
      - Autosomal recessive
      - Extra basic groups (e.g., NH_2)
      - COAL: **Cystine**, ornithine, arginine, lysine
   3. Mechanism: Dibasic amino acids excreted in urine
      - Cystine forms hexagonal crystals at high concentrations ("sixtine")

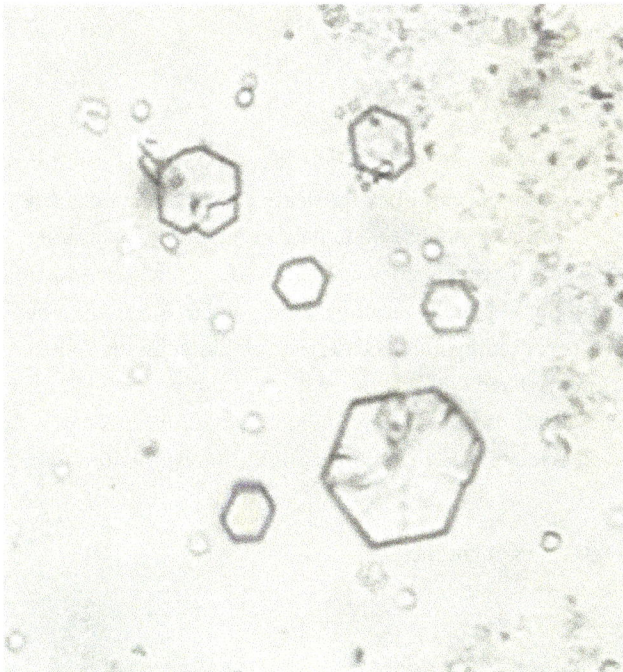

Figure 7-6. Cystine crystals. (Reproduced, with permission, from Kasper DL, Fauci AS, Hauser SL, Longo DL, Jameson JL, Loscalzo J, eds. *Harrison's Principles of Internal Medicine*. 19th ed. New York: McGraw-Hill; 2015.)

4. Presentation: Most common cause of kidney stones in children
   - **Pain** in the side and back (unilateral)
   - **Hematuria**
   - Higher risk of infection
5. **Treatment:** Increase fluids; alkanization of urine minimizes precipitation; low sodium, protein diet

7. Disease relevance summary

a. Amino acids play many important roles in the body beyond protein synthesis
b. Most common disease tested from this chapter is **phenylketonuria**
   1. Specific mechanism tests understanding of essential amino acids, and amino acid metabolism
   2. Mechanisms extend to similar diseases also (MSUD, homocystinuria)
c. Topics discussed here will come up in other sections of medical school curriculum
   1. **Hyperammonemia:** Caused by urea cycle disorders, but also hepatic encephalopathy
   2. **Cystinuria:** One cause of kidney stones, all with similar clinical presentation
d. Concepts will be useful in thinking through many questions in biochemistry
   1. **Properties of amino acids** can help assess how mutations might affect enzyme function or severity of disease
   2. **Essential** vs. **nonessential** will help with nutrition and related questions (Chapter 9)
e. **Urea cycle** questions come up less commonly, but covers **hyperammonemia** which is clinically important
   1. Focus on key steps and diseases, like **OTC deficiency**, without memorizing the entire cycle

Practice questions

1. A husband and wife come to see a genetic counselor. They are concerned about the possibility of their child having sickle cell disease, a condition characterized by anemia, acute, intense bone pain and hemolysis, and increased risk of stroke. They each have a history of the disease in their extended families. The genetic counselor explains that more than one DNA mutation can lead to disease, which is caused by a defective hemoglobin subunit; and that the type of mutation is correlated with the severity of symptoms. Which of the following amino acid substitutions, (normal → mutant), if inherited from both parents, would be expected to lead to a more significant dysfunction of the hemoglobin subunit?

   A. Asparagine → Glutamine
   B. Glutamate → Valine
   C. Lysine → Histidine
   D. Phenylalanine → Tryptophan
   E. Valine → Isoleucine

2. A 31-year-old man asks his primary care physician for referral to a nutritionist, as he wants to switch to a vegan diet. The nutritionist recommends a diet combining various sources of vitamins and proteins, including soybeans. However, the patient is allergic to soy, and asks if supplements could be used instead. The nutritionist explains that a combination of foods, including oatmeal, lentils, and nuts can suffice, but intake must be carefully monitored. At the end of the visit, the nutritionist provides guidelines for recommended daily intake of the essential nutrients. In which of the following pairs of amino acids both are considered essential?

   A. Asparagine, glutamine
   B. Aspartate, glutamate
   C. Cysteine, leucine
   D. Glycine, proline
   E. Lysine, phenylalanine

3. A gravida 2, para 1 woman has given birth to her first child; the other child died in utero. Her newborn daughter has a low birth weight and low Apgar scores, concerning the obstetrician. The newborn is taken to the neonatal ICU for observation. After testing the physicians make a diagnosis, and inform the parents that the disease cannot be cured, but it can be managed by feeding the child a diet rich in lipids and low in carbohydrates. What is the most likely mutation responsible for this patient's condition?

   A. Glucose-6-phosphatase
   B. Hexosaminidase A
   C. Hypoxanthine-guanine phosphoribosyltransferase
   D. Phosphofructokinase-1
   E. Pyruvate dehydrogenase

4. A 62-year-old man is brought to the urgent care clinic during the holidays by his daughter. She insisted he come when, while cooking dinner, his hand slipped and he cut his hand with a sharp knife. After you examine the injury, she comes out to tell you that she is worried about his worsening physical condition: he walks very slowly and seems to move with much more difficulty than last year. After conducting a neurologic exam, you suspect Parkinson's disease, a movement disorder resulting from decreased dopamine production in the substantia nigra. Interruption of which of the following enzymes could result in decreased levels of this neurotransmitter?

   A. Branched-chain α-ketoacid dehydrogenase
   B. Catechol-*O*-methyltransferase
   C. Dopamine β-hydroxylase
   D. Homocysteine methyltransferase
   E. Tyrosine hydroxylase

5. A homeless man is seen at a weekly free clinic with visible neurologic symptoms. He is highly irritable, and complains of decreased sensation in his hands and feet. He is able to give a partial history, and the medical student volunteer ascertains that consumes a substantial amount of alcohol and has unreliable access to a balanced diet. His symptoms are consistent with a nutritional deficiency, but

the medications he brought with him include thiamine, so the team considers options beyond a B_1 deficiency. If these symptoms are due to a deficiency of vitamin B_6, which of the following reactions would be inhibited?

A. Arginine → ornithine

B. Dopamine → norepinephrine

C. Glutamate → GABA

D. Phenylalanine → tyrosine

E. Serotonin → melatonin

6. A 2-year-old boy has undergone careful testing over the past year for delayed development, seizures, and lethargy. The symptoms are clearly related to hyper-ammonemia, but the physician has been unable to identify the source. Finally, consultation with a specialist and retesting of serum shortly after a meal identifies the issue as a mutation in an enzyme involved in catabolism of amino acids and excretion of nitrogen in the form of urea. What is the most likely lab result associated with this disease?

A. High serum citrulline

B. High serum orotic acid

C. High serum urea

D. Low serum sodium

E. Metabolic acidosis

7. A 3-year-old boy is brought to a daycare center by his parents. Both work long hours in a nearby factory after recently immigrating to the United States. The daycare manager soon becomes concerned that the boy appears to have several health issues. He has scaly skin and a mousy order, and does not appear to be learning English even after several weeks. The daycare manager suggests the child see a pediatrician for a checkup, and the pediatrician quickly suspects an enzymatic deficiency. Based on these symptoms, she recommends restricting a specific amino acid from the diet. In addition to this restriction, what supplement must be added to the patient's diet?

A. Biotin

B. Folate

C. Phenylalanine

D. Tyrosine

E. Urea

8. A 5-year-old girl was diagnosed with an enzyme deficiency after her parents noticed symptoms when she was an infant. She is now taking part in a study following patients with this condition longitudinally, to assess long-term symptoms and overall prognosis. These results are combined with those of participants with a variety of similar conditions to look for predictors of disease severity. Which of the following symptoms is associated with a relatively benign disease?

A. Body with a mousy smell

B. Elevated urine homocysteine levels

C. Elevated urine orotic acid levels

D. Urine that turns black in air

E. Urine with a maple syrup odor

9. A 28-year-old mother brings her week-old son to the emergency room after he suffers a series of seizures. Over the past few hours, he has grown lethargic and unresponsive. Although these symptoms arose over the past 12 hours, she did notice when changing his diaper that his urine had a sweet, syrupy smell, but did not think anything of it at the time. If this patient's condition is caused by an enzymatic deficiency, which of the following amino acids should be restricted from this patient's diet to manage his condition?

    A. Cysteine
    B. Isoleucine
    C. Lysine
    D. Phenylalanine
    E. Tyrosine

10. A 10-year-old girl comes in to see the school nurse during recess. While playing tag, she was suddenly overcome with severe pain in her side, which she says feels "like a knife." The school nurse tries to look at the patient but can barely get her to sit still, and calls her parents advising they go to the emergency room in case it is appendicitis. In the ER, a number of tests are done including ultrasound of the appendix, as well as a urine sample. The ultrasound does not suggest appendicitis, but the pathologist report notes hexagonal crystals in the urine, which the doctor suspects provides a diagnosis. What combination of interventions is most likely to improve the patient's condition?

	Fluid intake	Sodium	Urine pH	Protein
A.	↑	↑	↑	↑
B.	↑	↑	↑	↓
C.	↑	↓	↑	↓
D.	↑	↓	↓	↓
E.	↓	↑	↓	↑

Lipid Metabolism

CHAPTER 8

1. Purpose

a. Lipids make up the final big category of metabolism, in addition to carbohydrates and amino acids

b. Lipids are not only an efficient source of energy but also important for cell structure and signaling hormones

c. This chapter prioritizes topics like cholesterol and lipid transport, which are especially relevant to cardiovascular disease

2. Lipid synthesis

a. Recall acetyl-CoA from Chapter 5: Link between glycolysis and TCA cycle

b. Acetyl-CoA is also the branch point toward lipid synthesis

c. Overall, lipid synthesis can branch in 2 ways

    1. Fatty acid synthesis: Fatty acids (FAs), triglycerides (TGs) for fat storage

    2. Cholesterol synthesis: Important for cell membrane structure, steroid synthesis, and bile acid synthesis

d. Each of these relates to common topics in other systems, such as atherosclerosis, hypercholesterolemia, and diabetes mellitus

3. Fatty acid metabolism

a. Much lower yield than most of the other side pathways we'll talk about

b. Acetyl-CoA is the link between glycolysis and fatty acids

c. Key synthesis steps

    1. Branch point is at citrate—citrate is leaving mitochondria via citrate shuttle for cytoplasm where it:

      • Regulates glycolysis by inhibiting PFK-1 (prevent excess glycolysis)

      • Converts back into acetyl-CoA for fatty acid synthesis

    2. Biotin (B_7) is required for synthesis

    3. Synthesis occurs mainly where you would expect: Liver, and high-fat tissues that store lots of energy (adipose, mammary glands)

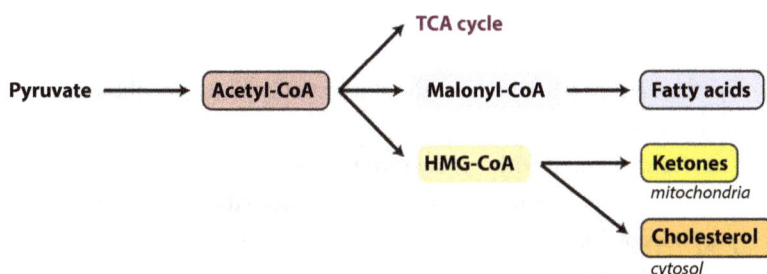

Figure 8-1. Overview of acetyl-CoA usage pathways.

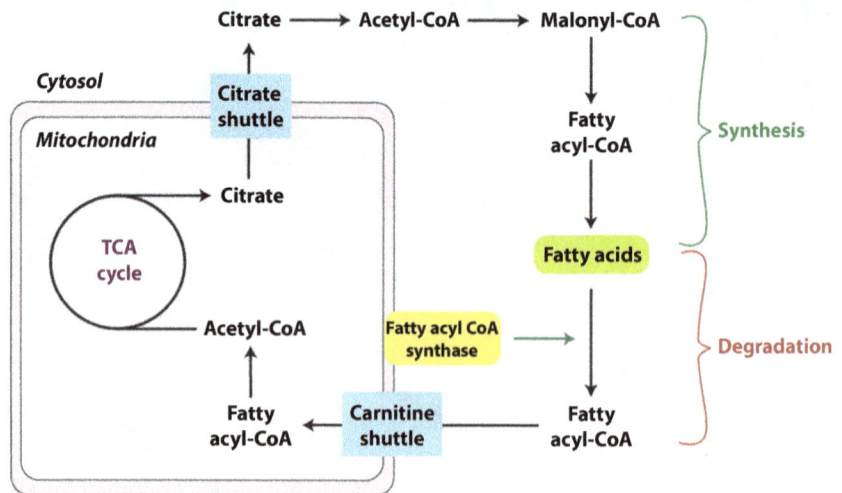

Figure 8-2. Fatty acid synthesis and breakdown.

d. Key breakdown steps
   1. Shuttle into the mitochondria to incorporate back into TCA cycle via carnitine shuttle
   2. Breakdown via β-oxidation to acetyl-CoA to enter directly into TCA cycle
   3. Even-numbered fatty acids cannot produce glucose (e.g., local energy source)
   4. Lipid metabolism in the liver is a major source of ketone bodies (see Chapter 9)
e. Clinical: Fatty acids (like glycogen) are a major energy source especially for muscle cells
   1. Systemic primary carnitine deficiency—can't get long chain fatty acids into the mitochondria leading to accumulation
      • **Weakness**, hypotonia—lack of energy, especially during fasting
      • **Hepatomegaly**—liver function is impaired
      • **Hypoketotic hypoglycemia**—liver can't process fatty acids for ketogenesis during fasting; and tissues that could have used fatty acids or ketone bodies for energy end up using up too much glucose
   2. Medium-chain acyl-CoA dehydrogenase deficiency—later in the pathway; can't break down 8 to 10 carbon FAs into acetyl-CoA
      • Similar symptoms (hypoketotic hypoglycemia) but roughly speaking, **more severe** (infancy/early childhood)
      • Additionally: Vomiting, neurologic symptoms (brain uses ketones under certain conditions), and liver dysfunction
      • Treatment: Avoid lipolysis by regular food intake

4. **Cholesterol synthesis**

   a. Synthesis takes place in the cytosol of virtually all tissues, though liver is a major source
   b. Carefully balanced with uptake through LDL receptors
   c. The cholesterol synthesis pathway can be distilled down to 3 steps for the purposes of Step 1
      1. Acetyl-CoA is converted into HMG-CoA to begin the process
         • HMG-CoA inside mitochondria is a precursor to ketogenesis, while in cytosol is a precursor to cholesterol synthesis

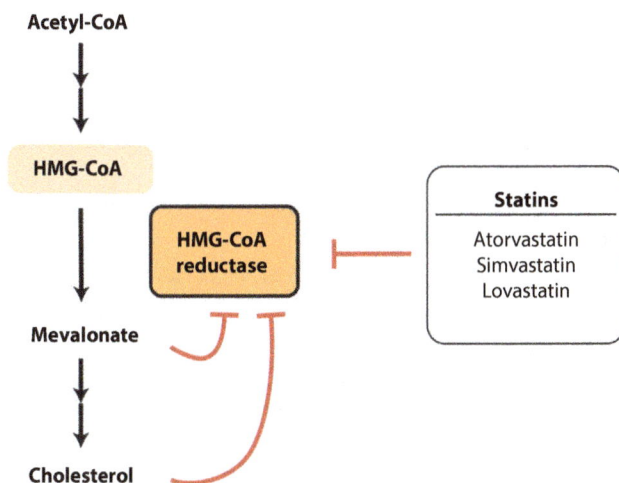

Figure 8-3. Cholesterol synthesis.

     2. **HMG-CoA reductase** converts HMG-CoA to mevalonate

     3. Many steps then lead to cholesterol (none important for Step 1)

  d. **HMG-CoA reductase** is the rate-limiting enzyme and is carefully regulated

     1. Feedback inhibition by mevalonate and cholesterol

     2. Inhibition by bile acids → cholesterol intake from diet via GI

  e. Cholesterol is used to synthesize many other compounds

     1. Steroid hormones (adrenal glands, gonads)

     2. **Bile acids**—absorb lipids from GI tract, and excrete excess cholesterol

       • Rate-limiting synthesis step: **Cholesterol 7α-hydroxylase**

  f. <u>Clinical:</u> Cholesterol has a large role in **cardiovascular disease** including **atherosclerosis**

     1. Cholesterol circulates in the blood in lipoprotein complexes (**LDL, HDL**)

     2. **Statin drugs** make up a major part of cardiovascular pharmacology

     3. Statins inhibit HMG-CoA reductase, reducing cholesterol synthesis in the liver

     4. Liver increases uptake of blood lipoprotein to maintain cholesterol stores → decreased blood lipid levels

5. Lipid transport

  a. Lipids are hydrophobic → must be packed for transport throughout the body

  b. Lipoprotein particles are bundles of TGs and cholesterol in a phospholipid particle with associated proteins

     1. Classified by density (and size): Low density = more fat (compared to protein) and larger

     2. Relative amounts of TGs and cholesterol vary depending on type and where lipids are being transported

  c. Types of lipoproteins

     1. **Chylomicrons** come from gut (high TGs) → drop off TGs in tissues → chylomicron remnants are collected by the liver

     2. **VLDL** (very low density) is similar to chylomicrons but comes from liver

     3. **IDL** (intermediate density) is similar to chylomicron remnants, returns leftovers of VLDL to liver

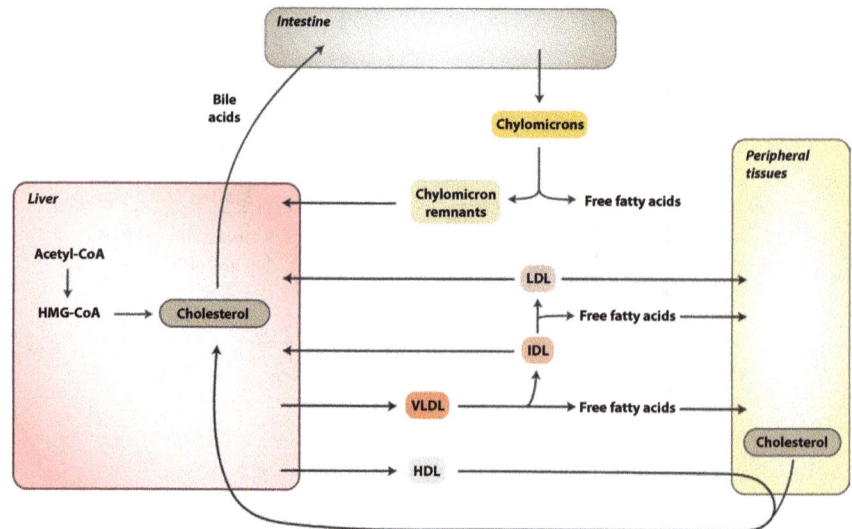

Figure 8-4. Lipid transport.

    4. LDL (low density) delivers cholesterol to tissues
- Associated with increased risk of coronary artery disease

    5. HDL (high density) returns cholesterol to liver
- Associated with decreased risk of coronary artery disease

  d. Key enzymes mediate lipoprotein synthesis and transport

    1. Lipases process TGs—remove from lipoproteins or intracellular stores
- Pancreatic
- Hepatic
- Lipoprotein lipase—processes TGs from particles into tissue; on vascular endothelium (expression regulated by PPAR-α)
- Hormone-sensitive—inside adipocytes, to release TGs

    2. Cholesterol ester modifying proteins affect packing of cholesterol into lipoproteins
- Lecithin-cholesterol acetyltransferase (LCAT)—esterifies plasma cholesterol → draw peripheral cholesterol into HDL via concentration gradient
- Cholesterol ester transfer protein (CETP)—transfers cholesterol esters from HDL to other particles → supports LCAT

Table 8-1. Lipoproteins.

Lipoprotein	Primary lipid	Source	Destination
Chylomicron	Triglycerides	Intestine	Peripheral tissues
Chylomicron remnant	Cholesterol	Chylomicrons	Liver
VLDL	Triglycerides	Liver	Peripheral tissues
IDL	Cholesterol, triglycerides	VLDL	Liver
LDL	Cholesterol	VLDL	Liver and peripheral tissues
HDL	Cholesterol	Liver, intestine	Liver

3. Apolipoproteins—signaling molecules help transfer lipoproteins around
   - A, B, C, E are important; E is the most important for Step 1
   - Can organize by letter to remember function

e. Clinical: Lipid transport is a major area tested on Step 1: LDL/HDL is closely associated with cardiovascular health, and lots of pharmacology
   1. This will come up across multiple topic areas as you study for Step 1
   2. Pharmacology: Several drug classes regulate lipid transport including statins, fibrates, niacin, and bile acid sequestrants
   3. Genetic deficiencies: Familial hypercholesterolemia and other dyslipidemias are due to mutations in lipid transport proteins
   4. Genetic risk factors: ApoE isoforms affect risk of Alzheimer's disease
      - **ApoE2** decreases risk
      - **ApoE4** increases risk

f. Pharmacology
   1. Statins (atorvastatin, lovastatin, simvastatin)
      - Inhibit HMG-CoA reductase → less cholesterol synthesized by liver → more LDL receptors to import from periphery
   2. Fibrates (fenofibrate, gemfibrozil)
      - Activate PPAR-α → increase LPL synthesis → decrease TG levels; increases HDL production
   3. Niacin (vitamin B_3; NAD^+)
      - Raises HDL, lowers VLDL → decrease lipolysis, TG transmission to liver → decrease VLDL production
   4. Bile sequestrants (cholestyramine, colesevelam)
      - Bind bile acids in GI → prevent reabsorption, increasing cholesterol excretion

Table 8-2. Apolipoproteins.

Apolipoprotein	Key lipoprotein	Function	Effect
ApoA-I	Chylomicrons, HDL	Activate LCAT	Uptake plasma cholesterol
ApoB-48	Chylomicrons	Secretion into circulation	Proper uptake
ApoB-100	VLDL, IDL, LDL	Bind LDL receptor	Uptake into liver
ApoC-II	Chylomicrons, VLDL, HDL	LPL cofactor	Cleave triglycerides
ApoE	All *except* LDL	Bind LDL receptor	Remnant uptake

Table 8-3. Lipid-lowering drugs.

Class	Mechanism	Use	Side effects
Statin (atorvastatin, lovastatin)	Inhibin HMG-CoA reductase	Lower LDL	Liver, muscle toxicity (rare)
Fibrate (fenofibrate, gemfibrozil)	Activate PPARα → LPL synth	Lower TG Raise HDL	Gallstones, myopathy
Niacin (vitamin B_3)	Decrease lipolysis in adipocytes	Raise HDL Lower LDL	Flushing, hyperuricemia
Bile sequestrant (cholestyramine, colesevelam)	Prevent reabsorption of cholesterol	Lower LDL	GI (nausea, cramps) Impaired vitamin uptake

Table 8-4. Genetic dyslipidemias.

Type	Affected protein	Effect	Clinical
I	ApoC-II LPL	Chylomicrons: ↑ TG ↑ Cholesterol	Pancreatitis, hepatosplenomegaly, pruritic xanthomas
II	ApoB-100 LDL receptor	*IIa:* LDL, cholesterol *IIb:* Also VLDL	Hypercholesterolemia → atherosclerosis, early MI; tendon xanthomas
III	ApoE	Chylomicrons, VLDL	Atherosclerosis, xanthomas
IV	Multiple	↑ VLDL ↑ TG	Pancreatitis

g. Genetic deficiencies
   1. Each associated with mutations in one of the major apolipoproteins or associated enzymes
   2. Associate protein function to type of lipid primarily affected
   3. Often associated with **xanthomas** (especially types I–III)
   4. <u>Type I:</u> ApoC-II or LPL
      - Chylomicrons increased
      - Hypertriglyceridemia, hypercholesterolemia
      - **Acute pancreatitis** (high TG levels → local lipase cleavage to fatty acids)
   5. <u>Type II:</u> ApoB-100 or LDL receptor
      - LDL increased
      - IIa: LDL only; IIb: LDL + VLDL
      - **Atherosclerosis**, early myocardial infarction
      - **Incomplete dominance:** Heterozygotes have intermediate phenotype
   6. <u>Type III:</u> ApoE
      - Chylomicrons, VLDL
      - **Atherosclerosis**
   7. <u>Type IV:</u> Multiple causes
      - VLDL production increased
      - Hypertriglyceridemia
      - **Acute pancreatitis**
h. ApoE and risk of Alzheimer's disease
   1. ApoE has three major isoforms: ApoE2, ApoE3, and ApoE4
   2. ApoE2 reduces risk of Alzheimer's disease
   3. ApoE4 increases risk of Alzheimer's disease
   4. Mechanism unknown; but thought to influence amyloid and Aβ secretion

6. Disease relevance summary

a. A lot of diseases brought up here because lipids closely relate to cardiovascular health
b. You will see a lot about cholesterol management and drugs when studying the cardiovascular system and pharmacology, so this is a good introduction

   c. The mechanism of **statins** as HMG-CoA reductase inhibitors and the effect on increasing liver LDL uptake is a key application of biochemistry

   d. Other lipid-lowering drugs have mechanisms that fit elsewhere in the pathway and have different effects on LDL, HDL and triglyceride levels

   e. The role of cholesterol in the synthesis of steroids and bile acids is important for reproduction and GI sections

   f. **Niacin** comes up in the next chapter on vitamins

   g. Come back to these concepts when working through questions in other organ systems

Practice questions

1. A 55-year-old man with type II diabetes mellitus takes metformin and long-lasting insulin, which he takes each morning when he wakes up. However, his primary care doctor is concerned because his blood glucose and HbA_{1c} are still above target limits even after 3 months of treatment. After testing fasting glucose levels throughout the day, the doctor prescribes a rapid-acting insulin to be taken before meals. Upon injection with the rapid-acting insulin, the rate of which of the following reactions would be expected to increase?

   A. Acetyl-CoA $\rightarrow$ Malonyl-CoA
   B. Fructose-1,6-bisphosphate $\rightarrow$ Fructose-6-phosphate
   C. Glucose-1-phosphate $\rightarrow$ Glucose-6-phosphate
   D. Linoleic acid $\rightarrow$ Acetyl-CoA
   E. Pyruvate $\rightarrow$ Oxaloacetate

2. A 3-year-old boy is brought to the pediatrician's office because of muscle weakness and loss of muscle tone. His parents have read about muscular dystrophy online, and are concerned that this is what he has. When the physician conducts a physical exam, however, she is confident that this is not his condition. Instead, she explains that there is most likely a deficiency of carnitine, a molecule in mitochondria. What is the pathophysiologic effect of carnitine deficiency?

   A. Decreased protein synthesis
   B. Defect in the electron transport chain
   C. Inability to synthesize ATP using the proton gradient
   D. Inability to break down long-chain fatty acids
   E. Unregulated activation of apoptosis

3. A 57-year-old woman presents to the primary care clinic for a routine checkup. Her medications include alendronate, aspirin, metformin, and simvastatin. Her physician notes that her LDL cholesterol (LDL-C) has dropped since her last visit, putting her within the target level <70 mg/dL. By which of the following mechanisms does this drug regimen most likely lower LDL-C levels in this patient?

   A. Activation of cholesterol 7α-hydroxylase
   B. Activation of PPARα
   C. Inhibition of HMG-CoA reductase
   D. Inhibition of lipolysis in adipocytes
   E. Inhibition of LDL receptor expression on hepatocytes

4. A 65-year-old man presents to the neurology clinic for evaluation of recent confusion and memory loss. His wife describes that he commonly has trouble finding his keys and wallet in the morning, even if he just put it down. Running a panel of cognitive functional tests, the neurologist suspects that the patient may be experiencing the initial stages of dementia. His wife notes that her husband's father developed similar symptoms later in life, leading the neurologist to suspect a possible genetic component to his disease course. The patient and his family are referred to a genetic counselor to discuss the possibility of genetic testing to assess risk for his children. Ultimately, however, the family decides not to pursue genotyping and the patient continues with a standard treatment regimen. Which of the following genetic variants, if present as a homozygous genotype, could be a contributing factor to the patient's condition?

A. ApoB-100
B. ApoC-II
C. ApoE2
D. ApoE4
E. LDL-R D69N

5. A 62-year-old woman has been taking medication for her cholesterol. The drug reduces cholesterol synthesis by the liver, increasing uptake of cholesterol from the blood. Some of this cholesterol is repackaged into lipoproteins and then released by the liver as needed. Which of the following lipoproteins does the liver release?

A. Chylomicron
B. LDL
C. VLDL
D. IDL
E. Chylomicron remnant

6. A 3-week-old baby girl is brought to the urgent care clinic over the weekend because of diarrhea and vomiting that occur after feeding. A blood test reveals that she is vitamin deficient and has almost no detectable lipids in the blood. The physician suspects a diagnosis of an extremely rare disorder in which lipid cannot be secreted from gut epithelial cells into systemic circulation. What protein is most likely deficient in this infant?

A. Apo A-I
B. Apo B-48
C. Apo B-100
D. Apo C-II
E. Apo E

7. A 44-year-old man comes into the clinic for his yearly checkup, and after taking a set of blood labs, the physician initiates a discussion about treatment for high cholesterol. He explains the transport of cholesterol in the body, including the intake of fatty acids into adipose tissue from circulating lipid particles. Which enzyme is responsible for this reaction?

A. Apo B-100
B. Apo E

C. HMG-CoA reductase

D. LDL receptor

E. Lipoprotein lipase

8. A pharmaceutical company is applying to the Food and Drug Administration for approval of a new drug to treat atherosclerosis. This drug induces synthesis of HDL by activating the PPAR-alpha transcription factors. What action of HDL is thought to make increasing plasma levels helpful in treating atherosclerosis?

A. Increases Apo E synthesis

B. Inhibits LDL production by liver

C. Promotes cholesterol secretion into the intestines

D. Reduces fatty acid uptake from the intestines

E. Removes cholesterol from tissues

9. A 12-year-old girl is admitted to the hospital due to severe stomach and abdominal pain. She reports that the pain is strongest in the right upper quadrant. Her parents are with her and state that the pain began rapidly during lunch at a sandwich shop near their house. None of the rest of the family experienced these symptoms, but all ate different meals. Serum amylase levels are ordered, and measurements taken a few hours after pain onset were 600 IU/L (normal <300 IU/L). In addition, physical exam notes hepatomegaly and the presence of yellowish lesions on the patient's back. Mutation of what protein would best explain this patient's constellation of symptoms?

A. ApoB-48

B. ApoB-100

C. ApoC-II

D. HMG-CoA reductase

E. LDL receptor

10. A 63-year-old man is following up with the endocrinology clinic due to poorly managed hyperlipidemia and hypercholesterolemia. He has tried a number of drugs, and although this has provided some benefit, his LDL and triglyceride levels are both >200. The endocrinologist considered adding an additional medication to his regimen, which will increase hepatic cholesterol 7α-hydroxylase activity. Which of the following drugs is most likely being considered?

A. Atorvastatin

B. Cholestyramine

C. Gemfibrozil

D. Metformin

E. Niacin

Nutrition

CHAPTER 9

1. Purpose

a. A large part of biochemistry, covered in the past few chapters, focuses on energy production and storage

b. There are a few aspects of metabolism that don't fit strictly in one of the previous chapters, and are presented here as a way to start integrating information

c. Beyond these pathways, nutrition also covers vitamins, which contain a lot of information

d. Vitamin questions tend to focus less on biochemical pathways and more on clinical symptoms

e. This is the most memorization-intensive chapter within biochemistry, so we will prioritize information as much as possible

2. Ketone bodies

a. Energy source derived from fatty acids and certain amino acids

b. Ketones are used primarily in brain and muscle

1. Muscle cells derive energy from ketones in addition to glucose, glycogen, and fat

2. Brain uses mainly glucose, but under certain conditions switches to ketones → minimal glycogen use

c. Not often tested on their own, but important background for related topics

d. Clinical: Diabetic ketoacidosis results from poorly controlled diabetes (insufficient insulin, unopposed glucagon)

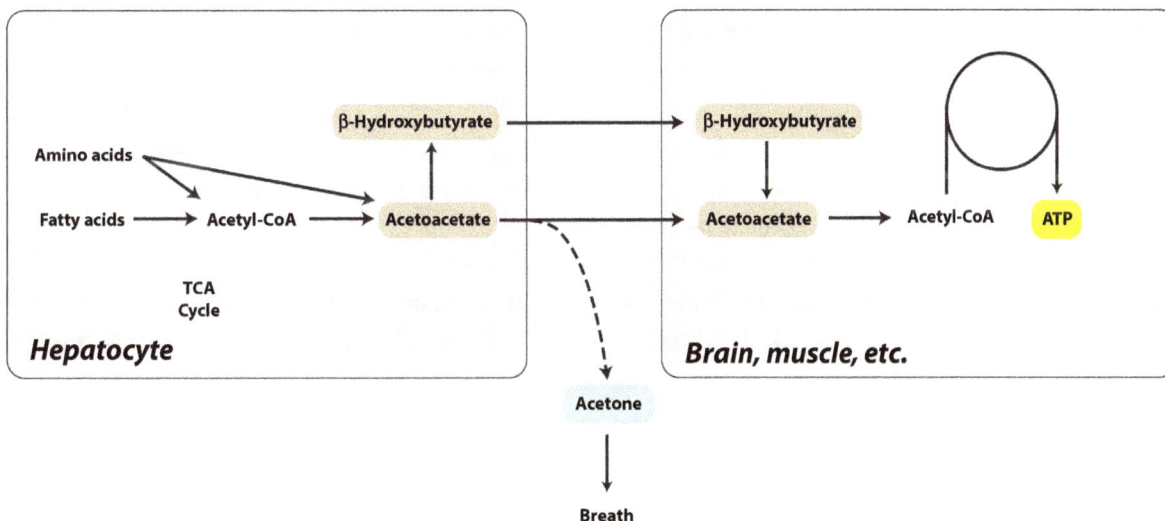

Figure 9-1. Ketone body metabolism.

1. Often first time diabetes is diagnosed: May recall previous polydipsia, polyuria (thirst and peeing)
2. Sudden onset of vomiting, tachypnea, confusion, progressing to loss of consciousness
3. **Metabolic acidosis:** Acidic ketones build up in the blood
4. Breath has a **fruity odor**: Acetone from some ketone bodies released through the lungs
5. Treatment: Insulin, hydration, and support

e. The ketogenic diet aims to reduce carbohydrates in favor of fats
1. In **pyruvate dehydrogenase deficiency,** pyruvate cannot be converted to acetyl-CoA
   - Shunt to lactic acid leads to acidosis with neurologic symptoms
   - Appears in infancy: **Failure to thrive**, developmental delay, **seizures**
2. In general, epilepsy disorders can benefit from a ketogenic diet
   - Possibly due to changes in neurotransmitter synthesis, lactic acid, etc.

3. Metabolism overview

a. We have now gone through the major arms of metabolism: Carbohydrates, protein, and lipids
b. Putting them all together is important for understanding how energy is used by the body in different contexts
c. Important for (1) daily activity and exercise and (2) starvation conditions
d. Exercise energy sources
1. Creatine phosphate
2. Anaerobic metabolism (glycolysis only)
3. Aerobic metabolism (using glucose, fatty acids, etc.)
e. Daily energy usage
1. Fed state: Glycolysis, aerobic respiration
2. Hours–1 day fasting: Glycogenolysis, gluconeogenesis, lipolysis
3. 2–3 days fasting: Adipose fatty acids, gluconeogenesis → muscle and liver fatty acids
4. Weeks fasting: Fatty acids → protein
5. Death: Protein degradation → organ failure

4. Protein-energy malnutrition

a. Severe acute malnutrition due to lack of protein and/or caloric intake
b. Common in childhood when needs are great, and in resource-limited countries especially
c. Classified based on presence of absence of *edema*
d. **Marasmus:** No edema; **very thin, muscle wasting**
1. Prolonged caloric deficit → muscle wasting and depleted fat stores cause emaciated appearance
2. Serum protein levels diminished, but not severely
3. Example in resource-rich countries: **Anorexia nervosa**
e. **Kwashiorkor: Edema;** peripheral edema, wasting; **central swelling**
1. Low levels of albumin → unable to hold fluids in vasculature leads to edema
2. Decreased apolipoprotein synthesis in liver: Swollen abdomen

3. Lack of protein synthesis → anemia, poor wound healing (skin lesions)

4. Examples in resource-rich countries: Fad diets or acute trauma/sepsis

f. <u>Treatment</u>: Careful treatment is necessary to avoid dangerous nutritional imbalances

5. Vitamins: Overview

a. Vitamins are obtained primarily through diet—so questions often focus on effects of deficiency (and occasionally of excess)

b. Grouped into fat-soluble or water-soluble because this affects when dietary deficiency can occur

1. <u>Fat-soluble</u> vitamin deficiencies associated with **malabsorption** especially of lipids
   - Steatorrhea
   - Crohn's disease
   - Bile acid sequestrants (e.g., cholestyramine)
2. <u>Water-soluble</u> vitamin deficiencies tend to result from specific diets
   - Folic acid (B_9): Lack of leafy greens
   - Thiamine (B_1): Alcoholism (most typically)

c. Within water-soluble, group vitamins by deficiency symptoms: Usually some combination of **dermatologic, neuro,** or **anemia** symptoms

d. Excess-related symptoms are less often tested but often easy to deduce based on deficits

e. Symptoms can be used to help remember vitamin function

6. Fat-soluble vitamins

a. There are 4 fat-soluble vitamins: A, D, E, K

b. Vitamin A: Retinol

1. <u>Function</u>: Cell differentiation; vision
   - Important for **retina** development → used to synthesize pigments in rods and cones
   - **Cell differentiation** → retinoic acid binds transcription factors to regulate cell growth

Table 9-1. Fat-soluble vitamins.

Vitamin	Key function	Deficiency	Clinical
A	Retina development; cell differentiation	**Night-blindness** Immune deficiency Epithelial symptoms	Acne treatment **Teratogenic**
D	Bone structure Calcium, phosphate absorption	**Rickets** in children **Osteomalacia** in adults	Kids: Bowed legs, deformities Adults: **Bone pain**, fractures
E	Antioxidant: Red blood cells, lipids	**Hemolytic anemia** Neuro: Ataxia, sensation loss	Involved in oxidation of LDL → increased atherosclerosis
K	Regulates clotting factors	**Anticoagulation**: Bleeding, bruising Risk ↑ in newborns, post-antibiotics	↑ PT, normal or ↑ PTT Newborns need vitamin K injection

      2. <u>Deficiency</u>: More common in resource-limited countries
- **Night blindness** (rods affected first)
- Immune deficiency → prone to infection
- Epithelial abnormalities: scaly skin; keratinization on corneas (Bitot's spots)
- Keratomalacia on cornea can lead to permanent blindness

      3. <u>Clinical</u>: Provide vitamin A to children in affected areas
- **Isotretinoin** (Accutane) is prescribed for severe acne
- Side effects: **Birth defects** due to cell differentiation effects → female users of Isotretinoin/Accutane must be on birth control

  c. Vitamin D: Calciferol

      1. <u>Function</u>: Calcium, phosphate regulation; and bone structure
- Promotes **calcium, phosphate uptake** from GI
- Activates osteo<u>blasts</u> at low levels → bone formation
- Activates osteo<u>clasts</u> at high levels → bone resorption

      2. <u>Deficiency</u>: Lack of sunlight; malabsorption conditions
- Children: Bones are still growing (epiphyseal plates are open) → deficiency leads to **rickets**
  - a. Bone shape and structural deformities: **Bowing** of leg bones; **bone pain**; delayed bone growth
- Adults: Loss of calcium and phosphate content in bones → **osteomalacia**
  - a. **Bone pain,** increased risk of **fractures** (or low-density pseudofractures), hyperparathyroidism

      3. <u>Clinical</u>: Vitamin D activation pathway is also tested in general
- Activation of 25-OH D_3 → active 1,25-$(OH)_2$-D_3 by proximal convoluted tubule kidney cells
- Interaction with parathyroid hormone → necessary for activation and phosphate balance
- **Breastfeeding infants** recommended to receive vitamin D supplementation

  d. Vitamin E: Tocopherol

      1. Much less often tested than the other fat-soluble vitamins; deficiency is very rare

      2. <u>Function</u>: Antioxidant, particularly of lipids and **red blood cells (RBCs)**

      3. <u>Deficiency</u>: Always secondary to fat malabsorption syndrome
- Examples: Pancreatic insufficiency, cholestatic liver disease
- **Hemolytic anemia**—RBCs are very sensitive to free radicals
- **Spinal cord neuropathy:** Especially neuromuscular—ataxia, loss of sensation/proprioception (lipids important in neuronal function)

      4. <u>Clinical</u>: Role in maintaining LDL/VLDL
- Vitamin E deficiency shown in research settings to alter LDL particles → increased atherosclerosis
- However, treatment with vitamin E does *not* have cardiovascular benefit

  e. Vitamin K: "Koagulation"

      1. <u>Function</u>: Integral to **clotting factor** production
- Factors II, VII, IX, X → central arm of extrinsic/intrinsic pathways
- Proteins C, S → anticoagulants

      2. <u>Deficiency</u>: Problems with decreased coagulation
- Easy **bruising, mucosal bleeding,** hematuria

Table 9-2. Water-soluble vitamins.

Vitamin	Key function	Deficiency	Clinical
B_1 (Thiamine)	**Energy production:** Decarboxylation of α-ketoacids (glycolysis, TCA, etc.)	**Beriberi:** Peripheral neuropathy (dry); cardiomyopathy, edema (wet) **Wernicke-Korsakoff** syndrome (eye, neuro symptoms)	*Beriberi* → malnutrition *Wernicke-Korsakoff* → alcoholism—can lead to permanent neuro symptoms
B_2 (Riboflavin)	**Energy production:** FAD: TCA cycle/ETC	**Skin symptoms:** Rash/inflammation of tongue, corners of the mouth (cheilosis) Normocytic anemia	Causes: Anorexia, lactose intolerance Associated with other deficiencies, so hard to pinpoint
B_3 (Niacin)	**Energy production:** NAD: TCA cycle/ETC NADP: HMP shunt, oxidative burst	**Pellagra (3 Ds):** Dementia, diarrhea, dermatitis (photosensitive)	Corn-rich diets of SE Asia Alcoholism **Excess:** Treatment for cholesterol causes flushing
B_5 (Pantothenic acid)	**Lipids:** Fatty acid, steroid synthesis	Burning feet syndrome	Rare; seen in severe famine
B_6 (Pyridoxine)	**Amino acids:** Neurotransmitter synthesis Heme synthesis	**Microcytic anemia** (sideroblastic) Peripheral neuropathy, **seizures** Dermatitis	**Drug-induced:** Isoniazid for TB Oral contraceptives L-dopa for Parkinson's
B_7 (Biotin)	**Carboxylation:** Gluconeogenesis Amino acid entry into TCA Fatty acid synthesis	**Rashes,** paresthesias	Avidin in **raw egg whites** prevents uptake
B_9 (Folate)	**DNA synthesis** (as THF) Purines dUMP → dTMP	**Megaloblastic anemia**	dTMP production is target of **multiple drugs:** Methotrexate (cancer) Trimethoprim, pyrimethamine (antimicrobial)
B_{12} (Cobalamin)	**DNA synthesis** (regenerates THF) **Myelination** (required for fatty acid processing)	**Megaloblastic anemia** **Paresthesias,** sensory and proprioceptive (gait) symptoms → can become permanent	Requires **intrinsic factor:** Deficiency in **pernicious anemia** (affects parietal cells) Ileocecal resection in **Crohn's disease** can cause poor uptake
C (Ascorbic acid)	**Collagen synthesis** Antioxidant	**Scurvy:** Easy bruising	Appears like **vasculitis** Too much can cause **GI distress**, iron excess, kidney stones

3. Clinical: Synthesized by gut flora
   - Deficiency seen in **neonates** (no intestinal flora) → give vitamin K injection at birth
   - Adults may have deficiency after **prolonged antibiotic use** (wipes out normal gut microbiome)

7. Water-soluble

a. In addition to vitamin C, there are 8 B vitamins tested on Step 1, for which the deficiencies may look similar

b. We try to group them within 3 categories based on symptoms, with most vitamins affecting multiple categories
   1. Skin—tends to relate to glucose catabolism
   2. Anemia—functions in DNA synthesis or as antioxidants
   3. Neuro—most severe deficiencies; centrally involved in pathways across metabolism, including lipid synthesis

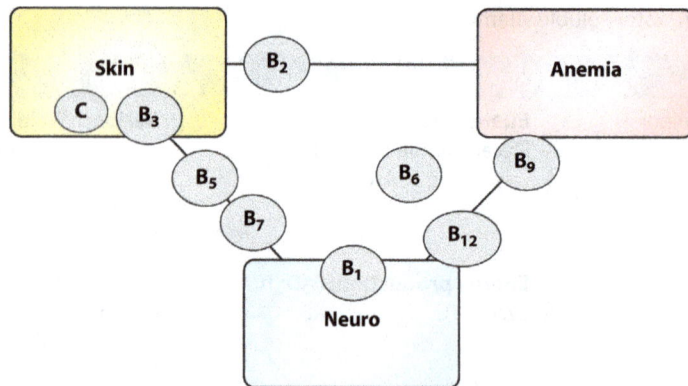

Figure 9-2. Clinical effects of water-soluble vitamins.

   c. Focus on deficiencies; knowing specific enzymes associated is lower yield than clinical symptoms

   d. Vitamin B_1: Thiamine

      1. Function: **Energy production** (ATP) pathways

- Think "glucose-related" → glycolysis, TCA cycle, pentose phosphate pathway
- Decarboxylation of α-ketoacids (changing number of carbon atoms)

      2. Deficiency: Rare: most commonly due to malnutrition or **chronic alcoholism**

- Malnutrition: leads to beriberi, disorder of muscle and nerves (paresthesias, pain)
  - a. Wet beriberi: **Dilated cardiomyopathy** → heart failure, peripheral edema
  - b. Dry beriberi: **Peripheral neuropathy** dominates; weak muscles especially in the legs (difficulty standing)
- Alcoholism: **Wernicke-Korsakoff**, similar combination of muscle/ neuro
  - a. Wernicke's encephalopathy: Ophthalmic symptoms → ophthalmoplegia, nystagmus; **confusion**, **ataxia**
  - b. Korsakoff syndrome: Chronic development of Wernicke's → **short-term memory loss**; confabulation

      3. Clinical: Maternal deficiency can lead to infantile beriberi; supplementation may be necessary (like vitamin D)

   e. Vitamin B_2: Riboflavin

      1. Function: **Energy production** pathways

- Redox reactions
- Component of FMN/FAD → electron acceptor in TCA cycle/ electron transport chain (~2 ATP)

      2. Deficiency: Mild; may go undiagnosed

- **Skin symptoms**, especially mucous membranes → cheilitis, sore throat, tongue inflammation ("magenta tongue")
- Can also see normocytic anemia
- Often associated with other vitamin deficiencies, so exact symptoms specific to B_2 aren't fully defined

3. <u>Clinical</u>: Situations related to **malnutrition** can lead to deficiency, including
   - Anorexia nervosa
   - Lactose intolerance—dairy is an important source

f. Vitamin B$_3$: Niacin (nicotinamide)
   1. Can be synthesized from tryptophan if necessary (but inefficiently)
      - Requires B$_2$ and B$_6$ for synthesis ($2 \times 3 = 6$)
   2. <u>Function</u>: **Energy production** pathways
      - Redox reactions
      - Component of NAD $\rightarrow$ electron acceptors in **TCA cycle**/electron transport chain (~3 ATP)
      - Component of NADP $\rightarrow$ pentose phosphate pathway, oxidative burst, and steroid/fatty acid synthesis
   3. <u>Deficiency</u>: Pellagra, characterized by a classic triad
      - **Dementia**
      - **Diarrhea**
      - Photosensitive **dermatitis**—hyperpigmentation in sun-exposed areas
   4. <u>Clinical</u>: Causes include
      - Alcoholism
      - Corn-rich diets—particularly in China, India, and Africa (cooking method affects availability)
      - Hartnup disease: Tryptophan excreted in urine due to lack of renal, GI transporters
      - Niacin treatment for cholesterol $\rightarrow$ **flushing**, hyperglycemia

g. Vitamin B$_5$: Pantothenic acid
   1. Least commonly tested vitamin on Step 1
   2. <u>Function</u>: CoA synthesis
      - Involved in **fatty acid** and steroid production—think "acid"
   3. <u>Deficiency</u>: Very uncommon
      - **"Burning feet"** syndrome $\rightarrow$ paresthesias, dermatitis
      - GI discomfort—though this is nonspecific
   4. <u>Clinical</u>: Classic context is during famine and war

h. Vitamin B$_6$: Pyridoxine
   1. <u>Function</u>: **Amino acid-related** pathways
      - Synthesis of neurotransmitters (histamine, dopamine)
      - Cofactor for liver enzymes ALT, AST (transamination)
      - Heme synthesis (from glycine)
      - Think "pyridoxine" and "histamine" to remember
   2. <u>Deficiency</u>: Generally related to compounds synthesized from amino acids
      - Heme: **Microcytic anemia** (sideroblastic)
      - Neuro: Peripheral neuropathy, hyperirritability, **seizures**
      - Generalized skin symptoms (**dermatitis**)
   3. <u>Clinical</u>: Deficiency rare but can be drug-induced (due to structure)
      - Isoniazid (tuberculosis)
      - Oral contraceptives
      - L-dopa

i. Vitamin B$_7$: Biotin
   1. <u>Function</u>: **Carboxylation** reactions
      - Carboxylation reactions $\rightarrow$ change in number of carbon atoms
      - Which specific reactions are not high-yield

- Reactions generally relate to pyruvate and acetyl-CoA
- Involve TCA-cycle intermediates or fatty-acid synthesis

2. Deficiency: General B-vitamin symptoms
   - Skin: Rashes, scaly skin
   - Neuro: Paresthesias, confusion

3. Clinical: Raw egg whites contain avidin, which binds biotin and prevents its absorption in GI track
   - Very uncommon in practice
   - Classic example is bodybuilder consuming lots of raw egg whites

j. Vitamin B_9: Folate

1. One of the most commonly tested B vitamins on Step 1—**most common** B-vitamin deficiency in the United States

2. Found in dark leafy-green vegetables

3. Function: **DNA base synthesis**
   - Active form is tetrahydrofolate (THF), inactive is DHF
   - Transfers methyl groups
   - Purine synthesis (A, G)
   - Synthesis of thymidylate from uracil (think "dTMP" base with THF → DHF)
   - Required for synthesis of methionine from **homocystine**

4. Deficiency: Classically **megaloblastic anemia**
   - Deficient DNA synthesis → RBC precursors can't complete cell cycle
   - Cytoplasm continues to grow without division

5. Clinical: Numerous ways to test this!
   - Pregnancy: Deficiency can cause neural tube defects like spina bifida
   - Drug interactions: THF conversion is a target of several antimicrobials and anticancer drugs (see Chapter 2)
     a. **Metho**trexate in cancer
     b. Tri**meth**oprim, pyri**meth**amine in infections
     c. All have some human enzyme activity (methotrexate highest)
   - Labs: Homocysteine in blood increased in deficiency
     a. In B_{12} deficiency, methylmalonic acid (MMA)

k. Vitamin B_{12}: Cobalamin

1. Overall, most similar to folate (B_9) as pathways overlap

2. Function: **DNA synthesis, myelination**
   - DNA base synthesis: Vitamin B_{12} is required to generate THF
   - Myelination: Cofactor for methylmalonyl-CoA mutase, required for myelin synthesis

3. Deficiency: Some similarities with folate
   - **Megaloblastic anemia** due to DNA synthesis
   - Symmetric **paresthesias**, numbness, gait disturbances
   - "Subacute combined degeneration" of the spinal cord—both dorsal and lateral columns
   - Changes are permanent if left untreated for too long

4. Clinical: Tie-ins with folate, and GI
   - Versus folate deficiency, look for: More severe **neurologic symptoms**, and increased methylmalonic acid in blood
   - Cobalamin absorption requires (1) synthesis by gut microbes, (2) **intrinsic factor (IF)** from parietal cells, and (3) terminal ileum

- **Pernicious anemia:** Anti-IF antibodies cause **B$_{12}$ deficiency** as well as **gastritis**
- Loss of ileum as in **Crohn's disease** (ileocecal resection)
- Found in animal products → diet-induced deficiency in veganism, though liver stores last several years

l. **Vitamin C: Ascorbic acid**
  1. Discussed also in Chapter 4
  2. Function: **Collagen synthesis**
     - Hydroxylation of proline and lysine allows crosslinking
     - Other functions: Important for **iron absorption**
     - Norepinephrine synthesis from dopamine
  3. Deficiency: **Scurvy** is most commonly tested
     - Poor collagen synthesis → weak extracellular matrix
     - Easy bruising, poor wound healing, hemorrhages
     - Hair: Misshapen (coiled "corkscrews"), perifollicular hemorrhage
  4. Clinical: Scurvy can resemble (misdiagnosed) as **vasculitis**
     - Excess intake can → iron excess, GI distress, and kidney stones

8. Disease relevance summary

a. Biochemistry plays important role in **balance and regulation of metabolism**
b. Vitamins especially have a lot of overlap with other organ systems: Heme/onc, GI → so review vitamins when studying those systems
c. Grouping vitamins by deficit symptoms helps
   1. Generalized B-vitamin deficit: **GI discomfort**, paresthesias, **dermatitis**
   2. B$_9$, B$_{12}$ (and B$_6$) share similar symptoms → pathway overlap
d. Save memorization for the weeks prior to the exam

Practice questions

1. A 14-year-old girl passes out during gym class at school and can't be woken up. An ambulance is called, and she is taken to the emergency room. It is discovered that she suffers from undiagnosed diabetes mellitus type I and is severely hyperglycemic. What is the most likely cause of her loss of consciousness?

   A. Hyponatremia due to high urine output
   B. Hypervolemia leading to high blood pressure
   C. Lack of glucose delivery to the brain
   D. Metabolism of fatty acids leading to acidosis
   E. Urea toxicity due to amino acid metabolism

2. A bodybuilder begins a strenuous workout. Shortly after starting his lifting routine, he feels a burning sensation in his muscles upon exertion that lessens when he rests between sets. He feels a similar sensation when sprinting, but not when he goes for long-distance runs. This sensation results most directly from metabolism of which of the following compounds?

   A. Citrate
   B. Creatine phosphate
   C. Ketone bodies

D. Linoleic acid

E. Pyruvate

3. A medicine resident is spending a month abroad in a developing country as part of her elective rotations in the third year of her program. While there, she assists in a pediatric clinic. A 3-year-old boy is waiting to be seen due to a progressive skin rash that has spread from his torso to his groin and legs. The parents state that they have limited access to food, and the family mainly eats corn-based meals with scarce protein or vegetables. The resident is concerned about the possibility of chronic malnutrition in this child. Which of the following symptoms, if present, would be indicative of a diagnosis of marasmus?

A. Elevated serum albumin

B. Enlarged abdomen

C. Lower-extremity edema

D. Muscle wasting

E. Poor night vision

4. An 8-year-old girl is brought to the emergency room after she was found living on the streets with her mother, who brought her in in desperation. She says that she and her daughter have been homeless since she left her boyfriend 3 months ago, and she has been trying to find a new place to live ever since. However, her daughter began to develop a skin rash that would not go away and has had a cold for 3 weeks at least. Deficiency of which of the following vitamins is most likely responsible for these symptoms?

A. Vitamin A

B. Vitamin B_1

C. Vitamin B_3

D. Vitamin D

E. Vitamin E

5. A pediatrician travels to Russia as part of a medical mission trip. He treats a 2-year-old boy who has trouble walking due to malformation of his legs. The bones of his legs have developed in a bowed shape, and he has poor muscle tone. His parents also note that he has a history of seizures. The underlying cause of this patient's condition also leads to which other disease?

A. Achondroplasia

B. Osteoarthritis

C. Osteomalacia

D. Osteoporosis

E. Paget disease

6. A newborn is kept for several days in the hospital for monitoring after a premature birth. While she seems to be mostly healthy, with no signs of respiratory distress or other problems, sustained bleeding from the umbilicus was observed, and blood is also detected in her stool. Treatment with which of the following compounds will most likely help address this patient's condition?

A. Vitamin A

B. Vitamin B_1 (thiamine)

C. Vitamin B_{12} (cobalamin)

D. Vitamin D

E. Vitamin K

7. A homeless man is brought to the emergency room by police. He was arrested for aggressive behavior and public intoxication. Upon physical examination, the man is incoherent. A blood test indicates that he has a blood alcohol level of 0.12%. He is kept overnight, and in the morning, the physician performs another full exam. He still shows signs of intoxication, including ataxia, even though a blood test indicates he is sober and taking no other drugs. Treatment with which of the following compounds is most likely to address this patient's underlying condition?

A. Folate

B. Vitamin A

C. Vitamin B_1 (thiamine)

D. Vitamin B_2 (riboflavin)

E. Vitamin D

8. A 60-year-old homeless man is admitted to the Veteran's Administration (VA) hospital after he was found unconscious at a bus stop and could not be aroused. After providing supportive care, the patient regains consciousness, but is not able to answer questions. The emergency medicine physician is concerned about aspiration pneumonia, and orders a portable chest X-ray. Physical exam of the patient notes dullness to percussion of the lungs on the right lower side. In addition, head and extremity exams note significant bruising on both legs and arms, as well as gingivitis on both the top and bottom gums. What process is most likely impaired in this patient?

A. Histamine synthesis

B. Hydroxylation of proline

C. Reduction of NAD^+

D. Renal secretion of phosphate

E. Tetrahydrofolate synthesis

9. A 27-year-old woman presents to the free family clinic asking to be seen by an obstetrician. She recently immigrated to the United States, where her husband has been working as a store clerk for 2 years saving up money. She is gravida 1, para 1, though she notes that her first child was stillborn due to a developmental issue. She now believes she is pregnant again, and wants to make sure the child is healthy. Her last menstrual period was 5 weeks ago. A pregnancy test indicates that she is pregnant, and the obstetrician believes conception was less than 3 weeks ago. She provides the patient with a prenatal vitamin cocktail and recommends, when possible, supplementing her diet with vegetables such as beans and avocados. Given the patient's history and current treatment, what condition is the obstetrician concerned about?

A. Down syndrome

B. Neural tube defect

C. Pellagra

D. Osteomalacia

E. Subacute combined degeneration

10. A 28-year-old woman has been a strict vegan for the past 2 years. She takes a daily vitamin because she was told by her doctor that veganism can lead to nutritional deficiency. However, she makes an appointment with her physician because she has recently been getting lightheaded on her runs even though she is careful to eat beforehand. She has also been noticing tingling of her hands and feet in the mornings. If her symptoms are related to vitamin B_{12} deficiency, which of the following findings is most likely?

A. Ataxia
B. Dermatitis
C. Excessive bleeding
D. Iron deficiency
E. Megaloblastic anemia

CHAPTER 10

Genetics & Disease

1. Purpose

    a. Genetics alone is a relatively small part of Step 1 but shows up as part of questions in almost every organ system

    b. The cell cycle provides an important foundation to discuss genetics, as cell division errors lead to diseases like Down syndrome

    c. Here we'll review some important concepts that can get easy points on Step 1

    d. We also review important genetic diseases seen in previous chapters to highlight how genetics can show up in many organ systems

    e. Lysosomal storage diseases represent a final review of genetics and review of concepts discussed previously

2. Cell cycle

    a. The cell cycle describes the common process of cell growth

    b. Phases can be divided into 3 main groups

    c. DNA Synthesis (S)

        1. The initial step toward cell division is replication of the DNA

        2. Each of the 23 pairs of chromosomes is duplicated, forming sister chromatids

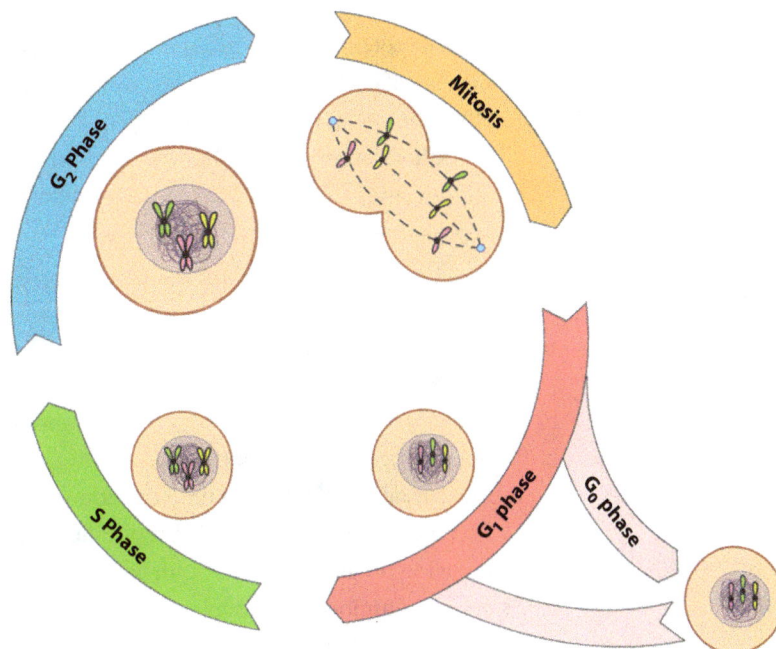

Figure 10-1. Overview of the cell cycle.

    d. Mitosis/meiosis (M)
1. After DNA synthesis and a period of rest, cells undergo division, and sister chromatids split to provide DNA for both daughter cells
2. Germ cells meant for sexual reproduction undergo meiosis to produce gametes (sperm or ova)
3. All other cells are somatic cells and undergo mitosis

    e. Gap/rest (G)
1. G_1: After recent division, the cell rests and carries out normal function
2. G_0: At this point the cell may enter a long-term rest phase if it does not need to regularly multiply
3. G_2: Between S and M phases, the cell grows and prepares organelles to support 2 new cells
4. After M phase, the cell enters G_1 and restarts the cycle, or may enter G_0

3. Cell cycle regulation

a. The cell cycle is a crucial process that requires careful regulation
b. Loss of regulation is a key mechanism underlying cancer
c. There are 3 major types of regulation to discuss
1. Cyclins
2. Tumor suppressors
3. Proto-oncogenes

4. Cyclins

a. Checkpoints at each step in the cycle prevent unrestrained cell division
b. The 2 protein classes to know are **cyclins** and **cyclin-dependent kinases**
c. Cyclins are the "time keepers" of the cell cycle
1. Cyclin concentration changes depending on the phase of the cycle
2. High levels of active cyclins are required to activate the next step of cell division
d. Cyclin-dependent kinases (CDKs)
1. Kinase proteins that are inactive in their resting state
2. Binding of cyclins is required to activate CDKs and drive cell maturation
3. However, cyclin binding is not sufficient to activate CDKs: Phosphorylation of the CDK is also required
e. Phosphorylation of CDK-cyclin complexes allows for many checks on cell cycle progression

5. Tumor suppressors

a. Many proteins that regulate the cell cycle were discovered because they are inactivated in cancer cells
b. Tumor suppressors are a class of genes that work to **limit cell growth**
c. Tumor suppressors are important for
1. Preventing DNA synthesis until mutations are corrected
2. Preventing cell division if DNA replication introduced new mutations
3. Inducing programmed cell death when damage cannot be repaired
d. Examples of tumor suppressor
1. Rb: "Retinoblastoma" protein
   - Acts in the G_1/S checkpoint — inhibits progression to DNA synthesis

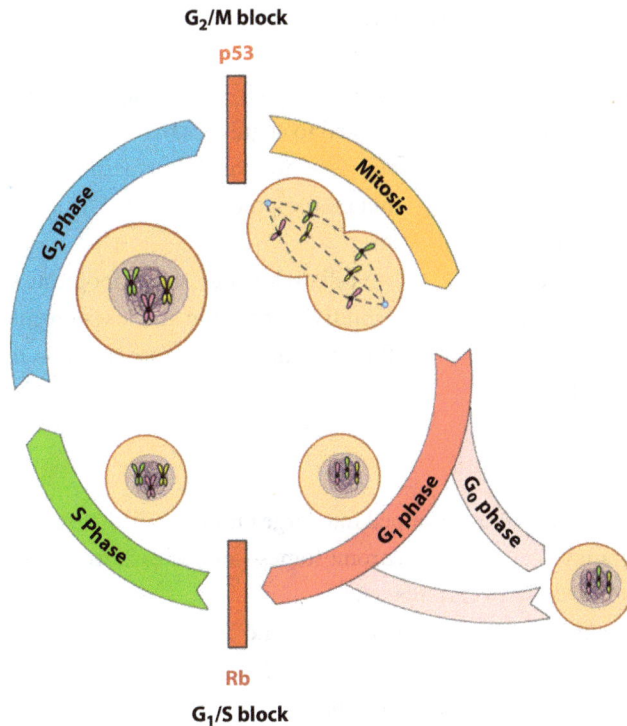

Figure 10-2. Regulation of the cell cycle.

- Rb activity is regulated by many signals, including growth factors (decrease activity) and cell differentiation (increase activity)
- Clinical: Retinoblastoma appears in the first years of life — shining light on the eye gives off an abnormal white reflection
- **Bilateral retinoblastoma** is always due to inherited Rb mutations

2. p53: Most common mutation in cancer
   - Acts in the **G_2/M checkpoint** — inhibits progression to mitosis
   - Levels increase when DNA damage is detected
   - Clinical: Li-Fraumeni syndrome — development of **multiple cancer types** early in life
   - Breast, bone, blood (leukemia), and more

   e. When both copies of the tumor suppressor gene are dysfunctional, the cell will not be able to detect DNA damage before progressing through the cell cycle

6. Proto-oncogenes

a. Proto-oncogenes are in many ways the opposite of tumor suppressor genes
b. When active, they **promote cell growth and division**
c. In normal physiology, they regulate their activity to activate growth only when necessary
d. Proto-oncogenes often act in growth factor signaling pathways, such as
   1. Receptor tyrosine kinases
   2. G-protein coupled receptors
   3. Transcription factors
e. Mutations in just 1 copy of the proto-oncogene can lead to oncogenesis

    f. Examples of proto-oncogenes
1. HER2/neu—growth factor receptor
- Receptor tyrosine kinase involved in growth factor signaling
- Activating mutations linked to aggressive breast cancer
- Trastuzumab (Herceptin)—antibody against HER2/neu protein — effective against the mutant receptor
2. Myc—transcription factor
- Normally plays a role in cell cycle progression, apoptosis
- Mutation leads to constitutive expression and loss of regulation
- Burkitt lymphoma: Rare but aggressive, more common in childhood

7. Chromosomes and genetics

a. Genes and DNA are grouped into large bundles called chromosomes
b. Humans have 23 pairs of chromosomes—one of each from the mother and father (total of 46 chromosomes)
c. One pair is the sex chromosomes, X and Y
1. Women have XX, and men have XY
2. Presence of the Y chromosome determines male sex
3. X is required for men as well because it is much larger and contains crucial genes
d. The remaining 22 pairs are called autosomal

8. Chromosomal abnormalities

a. Most genetic diseases we have seen result from mutations in DNA sequences causing errors in protein expression or function
b. However, some diseases, including Down syndrome, result from errors in chromosome organization
c. Aneuploidy describes an abnormal number of chromosomes; typically, one too many or one missing from the complete set of 46 in human diploid cells

Table 10-1. Sex chromosome aneuploidies.

Karyotype	Condition	Features
XO	Turner's	Stocky stature: Short; "shield chest" Lymphedema (seen in infancy) Ovarian dysgenesis ("streak ovary") Cardiovascular: Coarctation of the aorta, dissection Horseshoe kidney
XXX	Triple X	Most are phenotypically normal—a majority of cases are incidental findings Findings: Minor IQ deficits, tall stature, premature ovarian failure
XXY	Klinefelter's	Birth: Normal male Adolescence: Gynecomastia, infertility, tall stature Increased incidence of autism, psychiatric disorders
XYY	Double Y	Tall stature, developmental delay, requiring additional support Fertile; may not be specifically diagnosed

d. Nondisjunction is failure to separate—disjoin—chromosome pairs or sister chromatids
  1. Leads to aneuploidy with altered expression levels of proteins, causing disease
e. Most cases of aneuploidy are nonviable but individuals with specific chromosomal alterations may survive
f. Monosomy
  1. Turner syndrome is a monosomy of the X chromosome
    • Patients are female, but have some masculinizing features
    • Cause can either be meiotic or mitotic nondisjunction
g. Trisomy
  1. Occurs with only X/Y and chromosomes 13, 18, and 21
  2. Sex chromosomes: XXY (Klinefelter), XYY, XXX
    • Normal or nearly normal life span
  3. Autosomal trisomies
    • Patau syndrome (trisomy 13)
    • Edwards syndrome (trisomy 18)
      a. Both of the above are fatal in the first year of life
    • Down syndrome (trisomy 21)
      a. Viable into adulthood (~60 years)
      b. Many causes all lead to 3 copies of chromosome 21
    • Plasma protein levels can help detect in 1st–2nd trimesters
      a. β-hCG, PPAP-A, α-fetoprotein, inhibin A
      b. All down in trisomies except
      c. Down syndrome: β-hCG, inhibin A are increased

9. Mendelian inheritance

a. Mendelian inheritance refers to traits that are linked to specific genes and can be easily traced through offspring

Table 10-2. Autosomal trisomies.

Chromosome	Syndrome	Features
13	Patau	Rarest—often lethal before birth Holoprosencephaly, microcephaly, heart defects Polydactyly Increased risk of preeclampsia for the mother
18	Edwards	Often lethal before birth In utero: Growth retardation, polyhydramnios Heart defects (ventricular septal defect) GI: Omphalocele, Meckel diverticulum Severe intellectual disability
21	Down	1:500 pregnancies; increased in advanced maternal age Intellectual disability (esp. language production) Appearance: Flat facies, prominent epicanthal folds, flattened nasal bridge, transverse palmar crease Cardiovascular: Septal defects Ophthalmic: Myopia, astigmatism, Brushfield spots Dementia: Increased risk of Alzheimer's (encodes APP) Cancer: Increased risk of leukemia (ALL, AML)

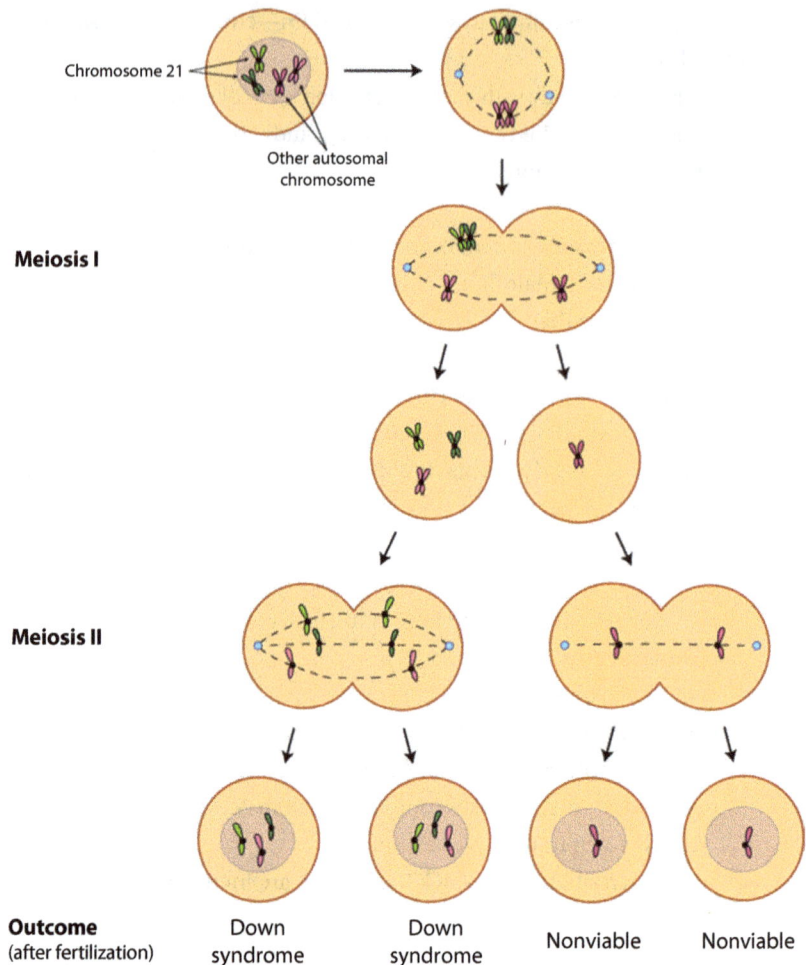

Chromosome 21

Other autosomal chromosome

Meiosis I

Meiosis II

Outcome (after fertilization)

Down syndrome   Down syndrome   Nonviable   Nonviable

Figure 10-3. Nondisjunction in meiosis I.

    b. Single gene (monogenic) inheritance is the simplest example, and many genetic diseases follow this pattern

    c. Different versions of the same gene are called alleles and can lead to expression of detectable traits called phenotypes

        1. Identifying phenotypes associated with disease allows for early detection and treatment

    d. Alleles can usually be classified as dominant or recessive

        1. Dominant: Presence of even 1 allele leads to a phenotype

        2. Recessive: Only expresses phenotype when dominant allele is absent (e.g., both copies of the gene have recessive allele)

10. Modes of inheritance

    a. **Pedigrees** are used to track genetic disease within the same family

        1. Investigating patterns of disease can help identify inheritance

          • Dominant vs. recessive

          • Autosomal or X-linked

          • Being able to quickly identify the most likely mode of inheritance from a pedigree will save a lot of time

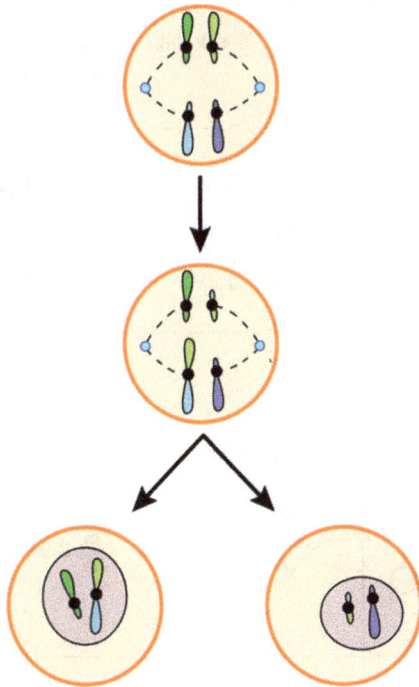

Figure 10-4. Robertsonian translocation.

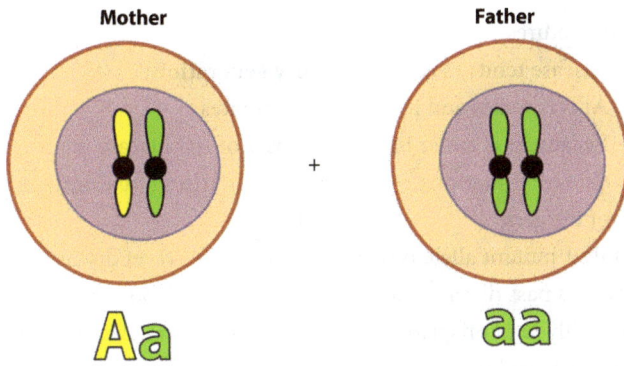

Figure 10-5. Mendelian inheritance cross.

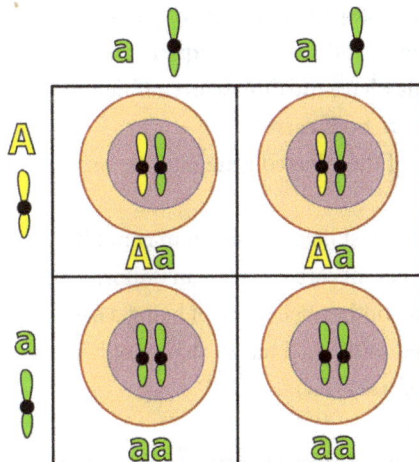

Figure 10-6. Punnett square.

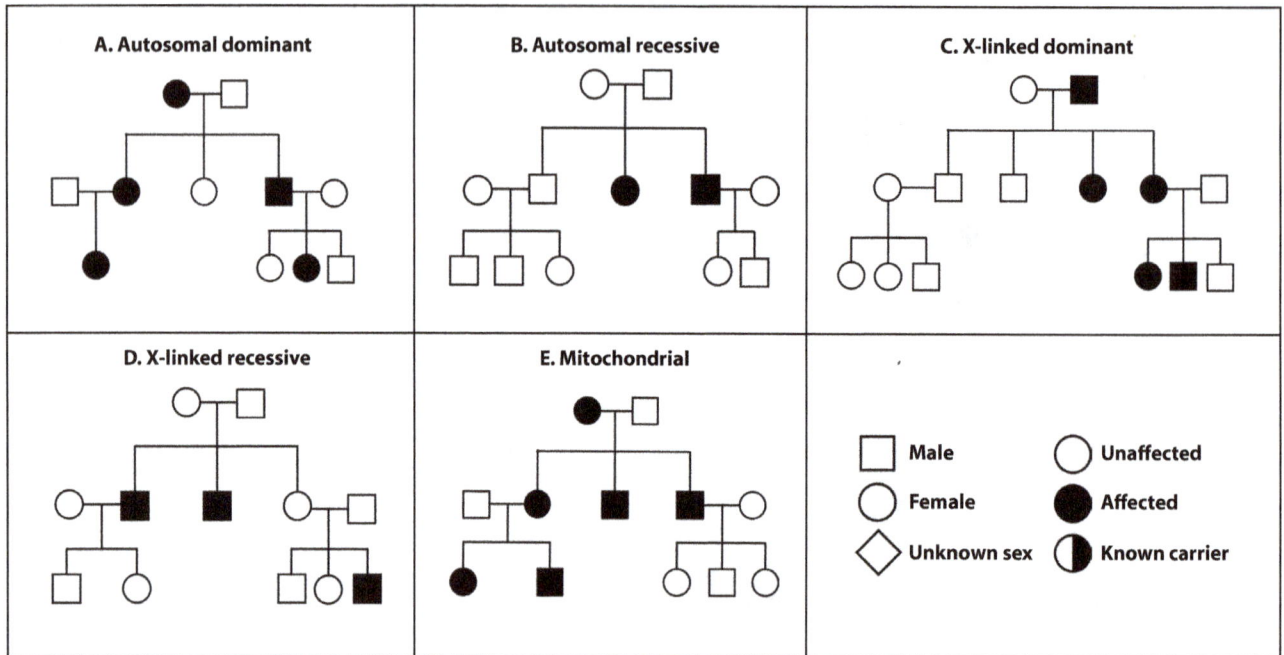

Figure 10-7. Modes of inheritance.

b. **Autosomal dominant (AD)**
   1. Key features
      - Disease tends to appear in **every generation**
      - Affects males and females at equal rates
      - Because it is easier to identify cases of AD disease, many arise from *de novo* mutations → new mutations arising randomly in only 1 copy of the gene
   2. Only 1 mutant allele is needed for a patient to get disease
   3. Parents pass disease alleles onto children with 50% probability
   4. Roughly ½ of offspring carry disease allele, and everyone that carries the allele will have the disease

c. **Autosomal recessive (AR)**
   1. Key features
      - Disease may **skip generations** (often there will be unaffected carriers)
      - Affects males and females at **equal rates**
      - A child may have the disease even if neither parent is affected (i.e., they are both carriers)
   2. Both alleles of the affected gene must be mutant for a patient to get disease
   3. Both parents must be carriers but may not have disease themselves
   4. When both parents are carriers, an unborn child has ¼ chance of being affected and another ½ chance of being a carrier; the unaffected sibling of an affected individual has ⅔ chance of being a carrier

d. **X-linked dominant (XD)**
   1. Key features
      - Disease appears in **every generation** → similar to AD
      - When mother is affected, the disease resembles autosomal dominant
      - When father is affected, all daughters and no sons are affected

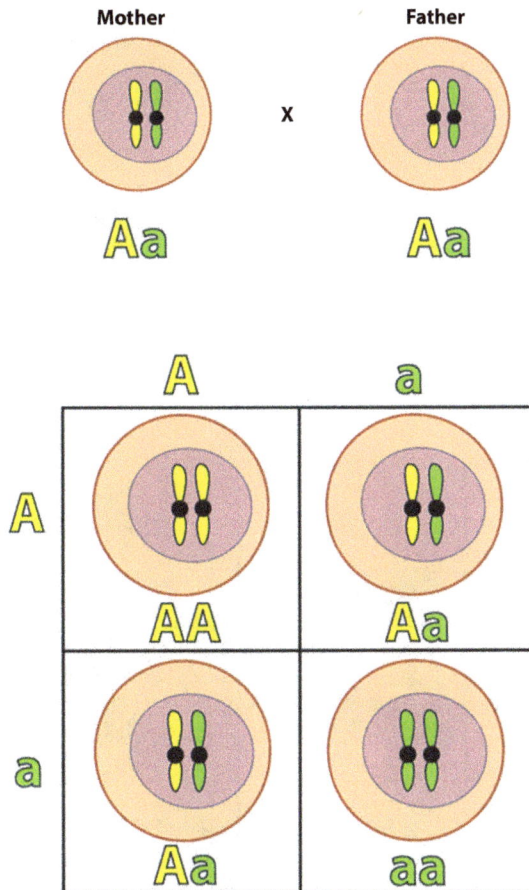

Figure 10-8. Heterozygote inheritance patterns.

2. Special inheritance pattern depending on whether disease allele is passed on from father or mother

3. Mothers pass on disease with 50% probability

4. Father only has one X chromosome, which is passed to all daughters and no sons

e. **X-linked recessive (XR)**

1. Key features

- Disease may **skip generations**
- **Males** are disproportionately affected
- All daughters of affected males are carriers, but no sons are affected (no male-to-male transmission)
- Roughly half of the sons of a carrier mother will have disease, while half of the daughters will also be carriers

2. Males only have one X chromosome, so have a 50% chance of getting the disease if their mothers are carriers

3. X-linked recessive disease has to be calculated differently for male and female offspring to account for number of X chromosomes

- Inheritance appears similar to AR in females, and uniquely in males
- **Hemizygous:** Affected males carry only the affected X chromosome and no others (the concepts of heterozygous and homozygous don't apply)

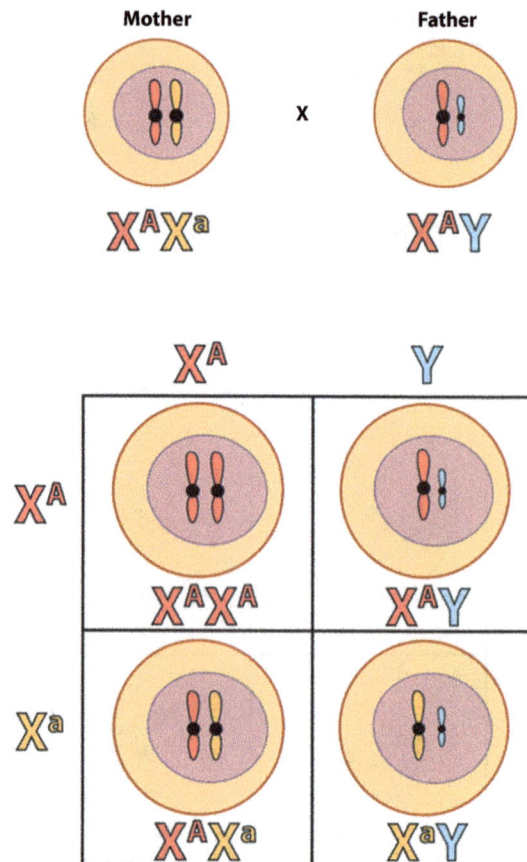

Figure 10-9. Sex chromosome inheritance.

    f. Mitochondrial
      1. Key features
- Affected **mothers pass disease** to all or nearly all offspring
- Affected fathers do not pass disease to any offspring

      2. Mitochondrial DNA is mutated
      3. Mothers contribute mitochondria to offspring, and as a result are the only ones who can pass on disease
      4. Most mitochondrial diseases have to do with **energy production** and involve **muscle** or **CNS symptoms** (high energy using systems)

11. Other factors influencing inheritance

    a. Deviations from Mendelian inheritance take 2 forms
      1. Expression of the mutant gene may be complex, and even if a patient carries the disease mutation, they may express disease differently
      2. Many diseases are polygenic and do not correlate to just 1 mutation
    b. Unusual inheritance
      1. Variable expressivity: The same disease can have **different presentation** in different people
- Example: Marfan's syndrome—classic symptoms (tall stature, eye and heart problems) are not present in 100% of patients

      2. Incomplete penetrance: A mutation may lead to **disease in some,** but not all people with that mutation
- Example: Familial cancer syndromes, such as BRCA1 and BRCA2 don't cause cancer in 100% of patients with the mutation

Table 10-3. Comparison of modes of inheritance.

Inheritance pattern	Identifying features	Clinical examples
Autosomal dominant	Every generation Men and women roughly equal	Familial cancer disorders (BRCA1, 2; Li-Fraumeni)
Autosomal recessive	Skips generations Men and women roughly equal	Loss-of-function Cystic fibrosis, phenylketonuria
X-linked dominant	All offspring of affected father 50% of offspring from mother	Extremely rare Cases mostly due to de novo mutations
X-linked recessive	Females rarely affected No offspring of affected father 50% of sons from carrier mother	Red-green color blindness Muscular dystrophy Lesch-Nyhan syndrome
Mitochondrial	From affected mothers only All offspring can be affected, (not always 100% = heteroplasmy)	Myopathies—lactic acidosis "Ragged red fibers" Examples: MELAS, MERRF

3. Codominance: For a given gene, **both alleles are expressed** even if only 1 copy is present
   - Example: Blood type is determined by presence/absence of A or B alleles: Presence of both → type AB blood
4. Locus heterogeneity: Mutations in **different genes** can lead to the same clinical disease
   - Example: Familiar hypercholesterolemia (Chapter 8) can be caused by mutations in different lipoproteins or associated proteins
5. Anticipation: Phenotype becomes more severe in each subsequent generation
   - Example: **Trinucleotide repeat disorders** such as Huntington's disease
   - Long strings of repeats cause even more replication errors, lengthening the repeat number → more severe symptoms

c. Imprinting
1. Imprinting is one of the more important processes modifying inheritance patterns to know
2. Although the child's genome has 2 copies of a gene, only 1 copy will be active
   - The gene "knows" if it is on the chromosome that was inherited from the father or mother
   - One is active as normal, while the other is permanently inactivated
3. If the copy that should be active is deleted, the other copy cannot take over
4. This may appear as a dominant genetic disease, though with the restriction that it depends from which parent the defective allele is inherited
5. Examples
   - Prader-Willi syndrome: Paternal copy is active normally, but mutated in disease; maternal copy is silent
     a. Symptoms: **Excessive eating**, obesity, **intellectual disability**
   - Angelman syndrome: Maternal copy is active normally, but mutated in disease; paternal copy is silent
     a. Symptoms: **Happy puppet syndrome—laughter**, seizures, **intellectual disability**

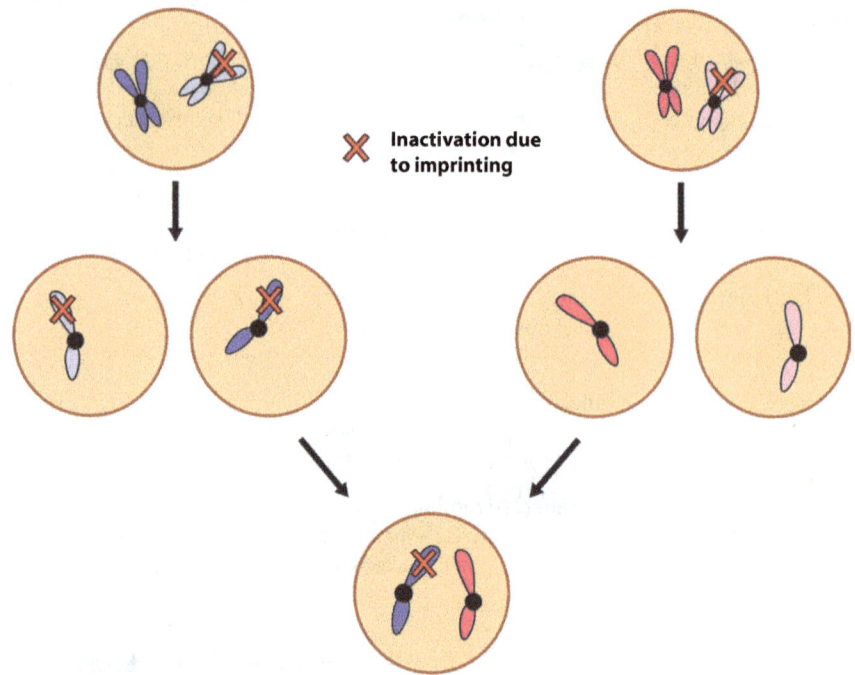

Figure 10-10. Imprinting.

    d. Polygenic disease
      1. Complex diseases can have a strong genetic component but many genes are involved
        • Schizophrenia, autism, inflammatory bowel disease, etc.
      2. Genetics interacts with environmental factors, so tracking disease is much more complicated

12. Monogenic disorders review

    a. Recall that a number of genetic diseases have been discussed in previous chapters
    b. These represent examples that are high yield for Step 1 for many reasons, but they can also come up in questions on genetics
    c. In addition to knowing the clinical presentation of the disease and biochemical effect, recalling the *inheritance* of the disease is useful
    d. Most diseases are autosomal recessive; next most common is X-linked recessive (more common in males)

13. Lysosomal storage diseases

    a. Lysosomes degrade many substances within the cell, including proteins, carbohydrates, and lipids
    b. Disorders of lysosomal enzymes interrupt breakdown and cause buildup of materials, which leads to cell toxicity
    c. All appear in **childhood** (most at infancy)
    d. For Step 1, Tay-Sachs is most common; knowing key symptoms and inheritance is useful; and enzyme if possible

Table 10-4. Review of monogenic diseases.

Disease	Inheritance	Effects	Chapter
Kartagener	AR	Ciliary dysfunction → respiratory problems, infertility	1
BRCA1 mutation	AD	Partial loss of homologous recombination enzyme → increased risk of cancer (breast)	2
Lesch-Nyhan syndrome	XR	Purine salvage deficiency → developmental delay, aggression	2
Sickle cell anemia	AR	Hemoglobin mutation → sickling, bone crises, autosplenectomy	3
Ehlers-Danlos	AD, AR	Collagen dysfunction → stretchy skin, aneurysms, joint dislocations	3
Marfan's syndrome	AD	Fibrillin mutation → tall, lens problems, risk of aortic dissection	4
Duchenne muscular dystrophy	XR	Dystrophin mutation → poor sarcomere function; progressive muscle weakness	4
Von Gierke	AR	Impaired glycogenolysis → hypoglycemia, hepatomegaly	6
G6PD deficiency	XR	Pentose phosphate pathway mutation → free radical damage to red blood cells with stress	6
Phenylketonuria	AR	Impaired phenylalanine metabolism → developmental delay, mousy odor	7
Cystinuria	AR	Impaired dibasic amino acid transport → cystine buildup in urine, kidney stones	7
Familial hypercholesterolemia, type II	Incomplete dominance	Mutated lipoprotein transport/receptors → elevated cholesterol, LDL	8
Tay-Sachs	AR	Lysosomal storage disease → neurodegeneration, cherry-red macula	10
Huntington's	AD	Trinucleotide repeat expansion → progressive neurodegeneration, chorea	10

Table 10-5. Lysosomal storage diseases.

Disease	Enzyme	Clinical
Tay-Sachs	Hexoaminidase-A	Cherry-red macula, neurodegeneration; no HSM
Niemann-Pick	Sphinomyelinase	Cherry-red macula, neurodegeneration; with HSM
Gaucher	Glucocerebrosidase	Bone crises, pancytopenia; "tissue paper macrophages"; with HSM
Fabry	α-galactosidase A	XR; peripheral neuropathy, angiokeratomas; later: CV, renal
Krabbe	Galactocerebrosidase	Peripheral neuropathy, optic atrophy, developmental delay
Hurler	α-ʟ-iduronidase (↑ heparan, dermatan sulfate)	Developmental delay, gargolyism, aggression; with HSM
Hunter	Iduronate sulfatase	XR; milder than Hurler

  e. Clinical: Common features are **neurodegeneration**, **cardiomyopathy**, and **hepatosplenomegaly (HSM)**

   1. Tay-Sachs: **Cherry-red macula**, neurodegeneration, no hepatosplenomegaly

   2. Niemann-Pick: **Cherry-red macula**, neurodegeneration, with hepatosplenomegaly

   3. Gaucher: Hepatosplenomegaly, **bone crises** with pancytopenia, "tissue paper macrophages"

   4. **Fabry: Peripheral neuropathy**, angiokeratomas, *later onset:* Renal, cardiovascular disease
   5. **Krabbe: Peripheral neuropathy**, optic atrophy
   6. **Hunter/Hurler:** Similar features with **gargoylism**, **aggression**, and **ophthalmologic symptoms** (retinal degeneration; corneal clouding)
   f. Treatment: Symptomatic; enzyme replacement therapy is available for some diseases

14. Population genetics and computation for Step 1

   a. There is an aspect of simple math to some genetics problems
      1. Probability of inheriting a disease from parents
      2. Likelihood that a parent is a carrier of disease given population frequency
      3. Use the Punnett square to aid in calculating from complicated pedigrees
      4. Probability trees can help bring together each step
   b. Examples of problems using Punnett squares
      1. Huntington's disease: Huntington's disease is an autosomal dominant neurological disease with an age of onset around 40–50 years old. A couple (both age 30) has just given birth to a daughter, when the father learns his own mother has been diagnosed with Huntington's disease. No one else in their families knows their status nor has a history of the disease. The family pedigree is thus shown in **Figure 10-11**. What is the chance this couple's newborn child has the Huntington's disease mutation?

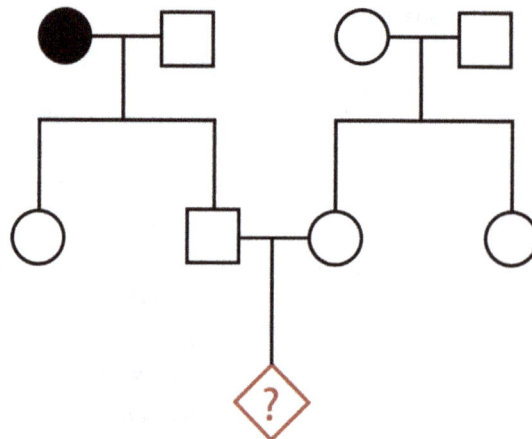

Figure 10-11. Huntington's disease pedigree.

   - Identify the people in the pedigree where you know their genetic status → the grandmother has a known mutation (Hh)
   - Identify the person whose status you want to calculate, and then which people through the pedigree must be calculated

   - The probability of passing on the mutant allele is ½ (50%)
   - This must occur twice → for the father, and then the newborn

$$\tfrac{1}{2} \times \tfrac{1}{2} = \tfrac{1}{4} \approx 25\%$$

2. Tay-Sachs: A couple is with a history of Tay-Sachs in their families is concerned about having a child with the disease. They do not know whether they are carriers. Their family pedigree is shown in **Figure 10-12**. What is the chance that the child indicated with the red diamond is affected?

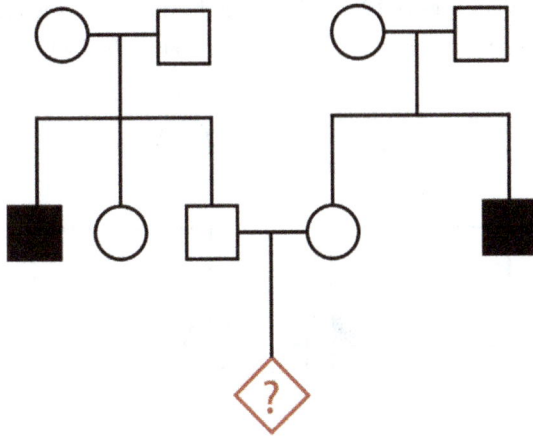

Figure 10-12. Tay-Sachs pedigree.

- We must calculate the probability that each parent is a carrier of the mutation

- The probability of each parent being a carrier is 2/3 because they are unaffected siblings of affected individuals → only 3 boxes of the Punnett square are possible
- What is the chance the child is affected if the parents are carriers?
- The child has a ¼ chance of having Tay-Sachs disease if both parents are carriers
- To calculate the final answer: **Multiply all probabilities** of events that must happen together

$$\tfrac{2}{3} \times \tfrac{2}{3} \times \tfrac{1}{4} = \tfrac{1}{9} \approx 11\%$$

c. Hardy-Weinberg equation for population genetics
   1. **Allele frequency** (p, q): Count all the copies of a gene in a population (2 × no. of people): How many are a given allele?
   2. Assumes random mating, no net loss or gain of mutations in the population
   3. Equations
      - $p^2 + 2pq + q^2 = 1$
      - $p + q = 1$
   4. Use these equations to calculate allele frequencies in the population

d. Example of Hardy-Weinberg genetics
1. <u>Phenylketonuria</u>: Phenylketonuria (PKU) is one of the most common inborn errors of metabolism, with a disease incidence in the United States of roughly 1 in 10,000 births. A couple hoping to have children is concerned because the father-to-be has a sibling with the disease and has been tested, showing that he is a carrier. The wife does not know her family history or carrier status. The pedigree for this family is shown in **Figure 10-13**. Based on the disease incidence, what is the estimated chance that she is also a carrier of the PKU mutation?

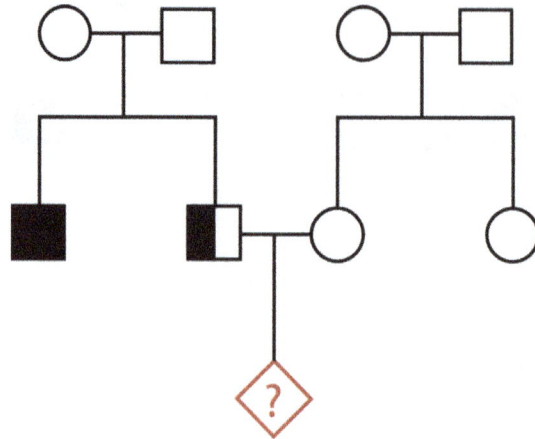

Figure 10-13. Phenylketonuria pedigree.

- What is the question asking? Only about the wife → no need to calculate the child's probability
- Disease incidence = $1/10,000 = q^2$

$$q = \sqrt{1/10,000} = 0.01$$
$$p = 1 - q = 0.99$$

- For rare disorders like PKU, it is safe to round p to 1

$$2pq = 2 \times 1 \times 0.01 = 0.02$$

- Use rounding when possible: The chance the wife is a carrier is about 2%
2. <u>Sickle cell disease</u>: An African American couple makes an appointment with the hospital's genetic counselor to discuss their plan to have a child. They are concerned because the wife has a history of sickle cell disease in her family: 2 siblings suffer from the disease, as does her mother, though she does not. The husband was adopted, and does not have access to his genetic history. The pedigree for this family is shown in **Figure 10-14**. Given this information, what is the chance the couple's child would have sickle cell anemia, if the allelic frequency of the sickle cell mutation in this population is 5%?
   - Take the question in parts. First, calculate the mother's carrier probability → whose status do we know for sure?
   - Her siblings are homozygous recessive, as is her mother: aa
   - No further calculation is necessary: The wife must be a carrier, Aa (we know she is not affected → cannot be aa)

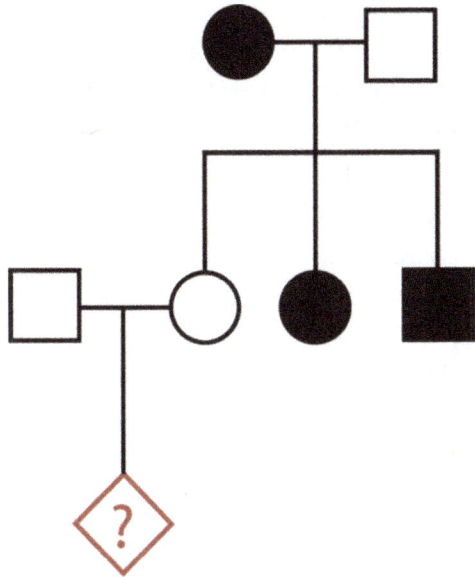

Figure 10-14. Sickle cell disease pedigree.

- How do we calculate the father's carrier frequency?
- "Allelic frequency" = q (sickle cell mutation)

- Knowing that $q = 0.05$, again round p to 1 to quickly find the proportion of carriers

$$2pq = 2 \times 1 \times (0.05) = 0.1$$

- To calculate the child's probability of having the disease, combine all of the factors that must occur together: (dad carrier × passes on allele) × (mom carrier × mom passes allele)

$$(0.1 \times 0.5) \times (1 \times 0.5) = 0.025 = 2.5\%$$

15. Approaches to studying genetics for Step 1

a. Cell cycle dysfunction is a common cause of inherited cancer syndromes
b. Errors in meiosis also underlie conditions like **Down syndrome**, which is commonly tested on Step 1
c. Finally, understanding concepts behind genetics and inheritance can provide easy points and save time for other challenging questions
d. However, only a selection of diseases are covered here—there are dozens covered on Step 1, and one of the best ways to learn them is to see them in practice questions as you go through question banks
e. Don't dwell on any one concept or term; genetics as a whole is a relatively small part of Step 1

Practice questions

1. A 2-year-old boy is brought to the pediatrician when his mother notices a strange photo-flash reflection in his right eye. She also worries about his vision,

as he often seems to be crossing his eyes and has difficulty making eye contact with people. The pediatrician performs an eye exam and identifies strabismus in the right eye, as well as leukocoria present in both eyes, but more strongly in the right eye. This patient likely has a mutation in a gene involved in which of the following cellular functions?

A. Chromatin condensation
B. $G_1 \rightarrow S$ cell cycle transition
C. $G_2 \rightarrow M$ cell cycle transition
D. Kinetochore formation in mitosis
E. Receptor tyrosine kinase signaling

2. A 62-year-old woman presents to the oncology clinic after diagnosis of liver cancer. She has been a chronic alcoholic for over 30 years, and the cancer diagnosis has caused her to recommit to getting treatment; she is now active in Alcoholics Anonymous and has been sober for 2 months. In analyzing her cancer, her physician analyzes a biopsy of cancer cells. She finds several mutations in both tumor suppressor genes and oncogenes. Which of the following is an example of an oncogene mutation?

A. IGF-1 receptor activating mutation promoting cell division
B. Impaired oxidative phosphorylation leading to anaerobic respiration
C. Loss of cyclin proteins allowing unrestricted cell growth
D. Mutation of DNA repair enzymes
E. Mutation of p53 inhibiting apoptosis

3. A 55-year-old woman notices on self-exam a lump in her left breast, and makes an appointment with her physician. The physician orders a mammogram and identifies a mass in the upper outer corner of the breast. Biopsy of the mass leads to a diagnosis of invasive ductal carcinoma. Further histologic analysis shows expression of a protein that makes this patient a candidate for treatment with the monoclonal antibody Herceptin (trastuzumab). The target of this drug is involved in which of the following processes?

A. Chromatin condensation
B. $G_1 \rightarrow S$ cell cycle transition
C. $G_2 \rightarrow M$ cell cycle transition
D. Kinetochore formation in mitosis
E. Receptor tyrosine kinase signaling

4. A couple in their 30s, expecting their second child, is called by the obstetrician with results of the amniocentesis. The results demonstrate decreased α-fetoprotein and PPAP-A, and increased β-hCG. The obstetrician recommends further testing, including karyotype analysis. Based on these tests, which of the following karyotype results would you expect?

A. 45,XO
B. 47,XXX
C. 47,XXY
D. 47,XX(+21)
E. 47,XY(+13)

5. A young couple who recently had their first child 6 months ago bring her to the urgent care clinic stating that she seems to have had a seizure. They live an hour

from the clinic, but note that they think this might be the second such event since she was born 6 months ago, and was more severe than the last time. On exam, the doctor notices several growths on the infant's cheeks, and a fibrous plaque on the forehead. When the physician points this out, the mother states that she has the same birthmark and it has never bothered her. After further discussion, the physician collects a more complete family history. He suspects that the pedigree for the condition leading to this child's symptoms is as follows, with the child shown in red:

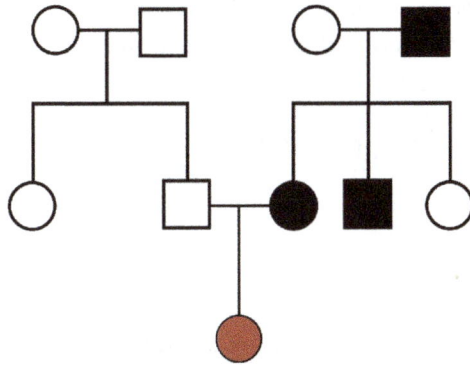

Based on this data, which is the most likely mode of inheritance for this condition?

A. Autosomal dominant
B. Autosomal recessive
C. Polygenic
D. X-linked dominant
E. X-linked recessive

6. A neurogenetics researcher has identified a small village in South America with a high incidence of an Alzheimer's-like disease, which she believes may be the result of a novel genetic mutation inducing high susceptibility. Interested in characterizing this condition further, she travels to the village and collects samples as well as the medical history of affected individuals and their family members. Once such family pedigree is shown below. The double line indicates a consanguineous (e.g., shared relatives) marriage.

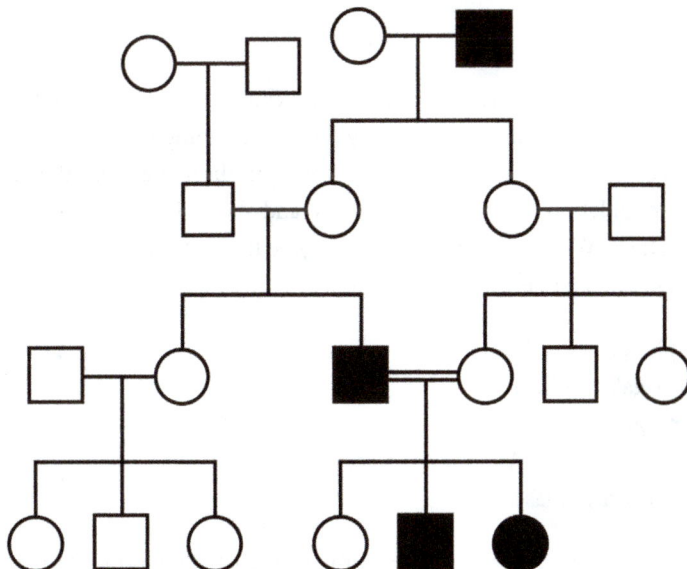

If these patients' dementia is the result of a single-gene mutation, what is the most likely mode of inheritance of the condition?

A. Autosomal dominant
B. Mitochondrial
C. Polygenic
D. X-linked dominant
E. X-linked recessive

7. A certain recessive disease affects 1 in 2,500 people in the United States. Assuming the requirements for Hardy-Weinberg are met, how many people are carriers in this population?

A. 1%
B. 2%
C. 4%
D. 8%
E. 20%

8. The pedigree shows family history of an autosomal recessive disease, which affects 1 in 400 people in the population. The chance that the child shaded in red has the disease is closest to which of the following?

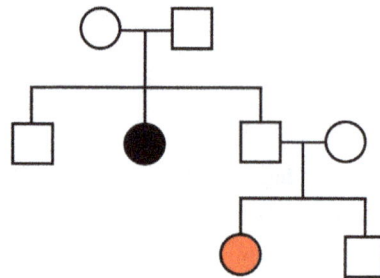

A. 1%
B. 1.5%
C. 2.5%
D. 6.3%
E. 10%

9. A new Jewish couple come to the genetics clinic. They would like to get married, but their parents insist that they receive screening to ensure that they are not carriers of Tay-Sachs, a relatively common disease among the Ashkenazi Jewish population. The geneticist agrees, and she collects DNA samples from each patient. What is the product of the gene that will be examined to test this couple's risk of passing on Tay-Sachs?

A. Alpha-galactosidase
B. Beta-galactosidase
C. Beta-glucosidase
D. Hexosaminidase A
E. Sphingomyelinase

10. A married couple makes an appointment with their family physician. They would like to begin a family, but want to consult with her beforehand. The

husband states that his sister had 2 children that both died less than a year after they were born. He says that they were diagnosed with Fabry disease, a lysosomal storage disease. The wife says her father's family also had the same disease. What should the physician tell this couple is the chance their child is affected with Fabry disease?

A. ~0%
B. 12.5%
C. 25%
D. 33%
E. 37.5%
F. 50%

11. A couple comes to the clinic with their 10-week-old son. They live in a rural area and had a home birth, so this is the first time their son has seen a physician. They are worried because, although their son developed normally at first, he has recently become very weak. The physician suspects a lysosomal storage disease, based on the presence of a cherry-red macula on ophthalmologic exam. What additional symptom would convincingly suggest the patient has Niemann-Pick disease?

A. Dermatitis
B. Hepatosplenomegaly
C. Mental retardation
D. Seizures
E. Skeletal deformations

Answers

Chapter 1

1. **(C) Nuclear receptor**

   Thyroid hormones enter the cell and signal directly within the nucleus. This is distinct from steroid hormones, which bind a receptor in the cytosol and are then transported to the nucleus; and protein hormones such as parathyroid hormone, which bind receptors on the plasma membrane and induce signaling within the cell without actually entering the cell. Because thyroid hormones bind their receptor directly in the nucleus, none of the other answer choices describe a step in the thyroid signaling pathway. [*Section 4.b*]

2. **(A) Diphenhydramine blocks neuronal signaling more readily than loratadine**

   First-generation antihistamines are nonpolar molecules, meaning they are not ionized and do not carry a charge, and they do not have a strong charge gradient. This means they can pass through cell membranes through diffusion relatively easily. Binding of antihistamines in the brain can lead to drowsiness, so first-generation antihistamines can have this side effect and are usually best taken at night before bed, when this will not be a concern. Second-generation antihistamines have a charged group on them, making them positively charged and unable to enter the brain through the blood brain barrier. None of the information contained in this question provides enough information to gauge the metabolism or effective dose of the drugs. [*Section 3.d*]

3. **(E) Luminal epithelial membrane**

   Lactose intolerance is caused by loss of the lactase enzyme. Lactase is present on the brush border (part of the apical polarized plasma membrane) of the gut epithelium, where it processes lactose from milk and dairy products into glucose and galactose (smaller molecules that are then imported into the cell). It is a commonly tested enzyme since loss of lactase is fairly common. [*Section 3.e*]

4. **(C) Glial fibrillary acidic protein**

   Glial fibrillary acidic protein, or GFAP, is an intermediate filament protein that is used to identify cell types in pathologic specimens. GFAP stains for astrocytes, which are one type of glial cells, or non-neuronal cells, in the brain (and the rest of the nervous system). This patient, who has an irregular mass in the temporal lobe on MRI, likely has developed a glioblastoma, but GFAP can also indicate less aggressive astrocytomas. [*Section 4.e, Table 1-1*]

5. **(F) Tubulin**

Several antimicrobial drugs target microtubule formation as their primary mechanism of action. Mebendazole is one example, used for parasitic infections (specifically, parasitic worms, known as "helminths"). These multicellular organisms rely on microtubule functions for many cellular functions, and the drugs are designed to specifically target parasite microtubules. [*Section 4.f*]

6. **(F) Endoplasmic reticulum**

Elevated liver enzymes indicate liver damage. Drug detoxification is one of the liver's major functions. In particular, the smooth endoplasmic reticulum is the site of detoxification, as well as steroid synthesis. This is in contrast to the rough endoplasmic reticulum, which classically is the site of protein synthesis. In this patient, hepatocytes are unable to cope with the amount of drug in the patient's system, and are dying (and releasing liver enzymes into the blood) as a result. This patient, experiencing opioid overdose, is given naloxone to block opioid receptors; but this will not address damage to the liver. [*Section 6.a*]

Chapter 2

1. **(C) It is relatively unmethylated and transcriptionally active**

Chromatin is the term used to describe DNA in its storage form within the nucleus. Chromatin is the combination of DNA wrapped around histone proteins and allowed DNA to be compacted into smaller, organized chromosomes. However, DNA existing in chromatin can still be accessed for transcription, by locally relaxing tight histone wrapping. DNA that is more loosely wrapped is called "euchromatin," which is less dense and thus appears lighter on imaging. Euchromatin is often more transcriptionally active, meaning genes in these regions are more highly expressed. DNA that is tightly wrapped in on itself is called "heterochromatin" and is generally inactive. Methylation is another level of epigenetic regulation of gene expression. Methyl groups can be added to the DNA directly (onto specific nucleotides) or onto histone proteins, both of which affect DNA wrapping and gene expression. In general, methylation of histones or DNA sequences decreases gene expression. On the other hand, acetylation of histones generally increases gene expression. [*Section 4.d*]

2. **(A) Methylation of the maternal chromosome**

Prader-Willi syndrome is a classic example of imprinting, another form of epigenetic regulation (regulation of gene expression beyond the DNA sequence itself). When a child gets half of his or her chromosomes from each parent, these chromosomes are not always interchangeable. In some cases, certain regions or genes on a chromosome are silenced when that chromosome comes from the mother, but not the father. That means, if the child's other copy of the chromosome contains a mutation at that site, disease could occur even though the sequence of one of the child's genes appears to be normal. In Prader-Willi syndrome, the paternal (father) gene is mutated, and because the maternal gene is silenced by methylation and, thus, the wild-type copy of the gene is not expressed, the child has the Prader-Willi phenotype. [*Section 4.f*]

3. **(B) Helicase**

   DNA replication involves several enzymes, but knowing general functions is usually sufficient for Step 1. This question asks about the enzyme typically responsible within the cell for enabling access of other replication enzymes to the DNA strand. Helicase, as its name may suggest, unwinds the DNA double helix strand to allow enzymes to access both single strands of DNA for simultaneous replication. Another enzyme, topoisomerase, plays a similar role by creating small nicks in the DNA strand to relax winding of the strand. However, topoisomerase is not listed here, so helicase is the best answer among the given choices. [*Section 5.b*]

4. **(D) Nonsense mutation at codon 39 of exon 2**

   DNA mutations are categorized by the specific type of mutation. A "silent mutation" refers to a change in the DNA code, replacing one base with another (a "point mutation"), that doesn't change the protein sequence coded by the DNA sequence. A missense mutation is a point mutation that does lead to a change in protein sequence, but it is hard to predict the severity of the mutation even knowing which amino acids are involved. Neither of these would affect the length of the associated protein (and in general, are categorized as less severe than the following mutations).

   A nonsense mutation changes a DNA base, turning a codon from one that encodes an amino acid to one that signals a stop codon, prematurely ending the gene and eliminating any regions of the gene after that. This would shorten the encoded protein. A frameshift mutation inserts or deletes DNA bases that leads to a shift in the reading frame, making all future codons different. However, introns are noncoding regions within a gene, meaning that mutations there are unlikely to affect the reading frame. Thus, choice D is the best answer. [*Section 10.c*]

5. **(E) Nucleotide excision repair**

   There are several different types of DNA repair mechanisms, depending on the source and severity of damage. This patient got a sunburn, which is caused by excessive exposure to ultraviolent (UV) rays in sunlight. UV rays are high-energy and can cause damage to DNA that distorts its structure. This is repaired by nucleotide excision repair, which removes the damaged section of DNA. This is in contrast to base excision repair, where a single base is mutated (bases are smaller than nucleotides). The first step in base excision repair is removal of the damaged base, whereas in nucleotide excision repair a whole portion of a single DNA strand is removed. [*Section 6.b*]

6. **(D) Phosphoribosyl pyrophosphate**

   This question is asking about *de novo* nucleotide synthesis, which is divided into purine (nucleotides A, G) and pyrimidine (nucleotides T, C) synthesis. The key point is that the researchers want to most effectively inhibit overall nucleotide production. Of the choices listed, phophoribosyl pyrophosphate (also called PRPP) is the molecule that is used in both purine and pyrimidine synthesis. Thus, a molecule designed to inhibit enzymes that normally use PRPP would be hypothesized to inhibit both pyrimidine and purine synthesis. [*Section 8.a*]

7. **(A) Accumulation of adenosine nucleotides**

This patient has SCID, which is commonly tested on Step 1 as one way to test purine salvage. Although there are many causes of SCID (involving many different cellular functions), one commonly tested cause is adenosine deaminase (ADA) deficiency. This enzyme is normally responsible for breakdown of adenosine into inosine, which is further broken down eventually into uric acid for excretion. The key is that if ADA is not present, adenosine accumulates. Cells carefully regulate levels of nucleotides and nucleosides, so this leads to dysregulation of DNA synthesis. Immune cells are highly proliferative and thus require a great deal of rapid DNA synthesis, so they are highly affected by this mutation, and failure of the immune system (specifically, lymphocytes) results. [*Section 9.e*]

8. **(E) Xanthine oxidase**

Gout is a common condition, particularly with age, that results from accumulation of uric acid in the body. These crystals infiltrate joints and precipitate as monosodium urate crystals, which causes inflammation as the body recognizes that these are out of place and mounts an immune response. The classic presentation is an older gentleman who eats a lot of red meat (which is relatively high in protein and purine nucleotides, which are broken down into uric acid), and in general, treatment aims to (1) reduce uric acid production and (2) increase uric acid excretion. A major source of uric acid production within the body is by purine breakdown, of which the final step is breakdown of xanthine to uric acid by xanthine oxidase. While some other choices listed are involved in purine breakdown, xanthine oxidase is the most downstream step and directly leads to uric acid production, making it the most direct target. Indeed, several drugs (such as allopurinol) treat gout by reducing production of uric acid. [*Section 9.e*]

9. **(B) Febuxostat**

These two patients are exhibiting aggressive (and possibly self-mutilating) behaviors as well as developmental delays, which are classic symptoms in young boys of Lesch-Nyhan syndrome. This disease is caused by a deficiency of HGPRT, which is involved in purine salvage. The enzyme is coded on the X chromosome, so girls (who have two X chromosomes) are much less likely to be affected. Deficiency of HGPRT leads to reduced recycling of purines, and thus more breakdown. This leads to high levels of uric acid, which can build up and cause severe gout and associated joint swelling, as well as kidney damage. The severe gout may be treated by limiting uric acid production by inhibiting xanthine oxidase. Febuxostat is a xanthine oxidase inhibitor, so is the best choice (the "xo" in the name can help remember **x**anthine **o**xidase). [*Section 9.e*]

10. **(D) 6-Mercaptopurine**

This question asks about nucleotide synthesis, which is a common target for anticancer therapy. Loss of cell cycle regulation is a hallmark of cancer, meaning cells divide much more rapidly than they normally should. This makes drugs that inhibit DNA synthesis, which is necessary for cell division, preferentially target cancer cells. Of the choices listed,

6-mercaptopurine is the only drug that inhibits *de novo* nucleotide synthesis. Many of the drugs in this category can be remembered based on their names, which can contain hints (such as 6-mercapto**purine** and 5-fluoro**uracil**). [*Section 8.e*]

Chapter 3

1. **(D) Promoter**

Many regulatory elements play a role in controlling gene expression. Some of these are at the protein level, such as transcription factors, and many are at the DNA level, in the form of specific sequences around the gene itself. Important regulatory DNA sequences to know are promoters, silencers, and enhancers. Promoters are sequences upstream (5′) of the gene that serve as the site of transcription factor binding, which initiates the transcription process. Promoters often feature specific sequences known as TATA or CAAT boxes. Promoters help drive cell-specific expression, allowing one gene to express differently depending on the needs of a particular cell. Silencers and enhancers also regulate gene expression, but they are less common than promoters. Promoters are the most reliable feature of the choices listed that would help drive cell-specific expression of a gene. [*Section 2.a*]

2. **(D)**

To determine the RNA sequence that would be generated by this DNA strand, the first step is to find the corresponding start sequence, denoted by the codon 5′ ATG 3′ (which becomes AUG in RNA). However, both strands should be examined as the DNA could be read from 5′ to 3′ on either the top or bottom strand.

A more efficient way to answer this question is to examine the answer choices and eliminate those that are visibly incorrect. Because the open reading frame of an RNA strand must begin with the start codon, (AUG), choices A and B are incorrect, having different codons. Choice C is incorrect because T is only found in DNA, so this is a DNA strand. Finally, of the remaining choices, the only difference is the presence of an additional codon. However, UGA represents a stop codon, meaning that the open reading frame would terminate at this site and not include the final UAG of choice E. Thus, D is correct. [*Section 4.b*]

3. **(B) 50S ribosome subunit**

Ribosomes are made of two subunits, named for the size they appear on molecular analysis. Prokaryotes (such as bacteria) and eukaryotes (such as humans) have different ribosome structures, making these good targets for antimicrobial drugs. Prokaryotic cells contain 30S and 50S subunits, while eukaryotic cells have 40S and 60S subunits (remembered by larger numbers for larger organisms). Both the 30S and 50S ribosome subunit are targets of certain classes of antimicrobial drugs, and inhibition of either leads to decreased protein synthesis and growth inhibition. Macrolides like erythromycin are one of the main classes that inhibin 50S, while aminoglycosides, like streptomycin, and tetracyclines inhibit 30S. For this question, 30S was not an option, so knowing the specific subunit target was not required, but it is important to remember that 30S/50S are prokaryotic. [*Section 4.d*]

4. **(D) Its catalytic elements are comprised of RNA**

Ribosomes are unique in major enzymes tested on Step 1 because they contain RNA that has a catalytic function. This goes against the typical view of enzymes as being proteins. Ribosomal RNA (rRNA) is the most abundant type of RNA within the typical cell (as compared to messenger RNA and transfer RNA). [*Section 3.a, Table 3-1*]

5. **(B) Poly-adenylation signal**

After initial transcription of DNA into messenger RNA (mRNA), several processing steps take place to stabilize mRNA and prepare for translation. This includes addition of a 7-methylguanosine cap on the 5′ (start) end of the RNA molecule, which helps stabilize the RNA. Throughout the molecule, introns are spliced out (removed) from the RNA sequence, and exons are joined together as the protein-coding region. Finally, at the end of the molecule, a poly-adenylation (-AAAA-) tail is added, which also helps stabilize the molecule and prepare it for export from the nucleus. The polyadenylation tail is ideal for reverse transcribing mRNA back into DNA for sequencing in the research setting because it is common to all mRNA molecules, but not other RNA molecules, and is long enough to target with a custom primer (similar to what is done in a polymerase chain reaction, or PCR). [*Section 3.b*]

6. **(B) Proline has a rigid structure that restricts proper alpha-helix formation**

Alpha-helices are secondary protein structures that involve tight turns in the primary protein structure; this allows amino acid variable (R) groups to face outward and maintain hydrophobicity or hydrophilicity (alpha-helical regions often serve as transmembrane domains due to the plasma membrane's lipid quality). Proline, and to a lesser extent, glycine, do not conform well to the alpha-helical structure. This is because proline has a circular R-group that exists at a rigid angle as cannot conform to the bend necessary for an alpha-helix. Glycine has a very small R-group (H) that makes it highly flexible and the resulting alpha-helix can become deformed. Thus, for proline, choice B is correct. (Choice A refers to cysteine, and choice E to glycine.) [*Section 5.c*]

7. **(B) A missense mutation in the hemoglobin beta-chain leads to an irreversible conformational change of hemoglobin in the deoxygenated state**

This patient has sickle cell anemia, a relatively common autosomal recessive disease that is more common in people of African descent. Hemoglobin is a protein in erythrocytes (red blood cells) that carries oxygen to tissues and carbon dioxide to the lungs for exhalation. It is made up of a quaternary structure of four subunits, two alpha and two beta, which must bind correctly together to create the functional protein. In sickle cell anemia, the beta-chain of hemoglobin is mutated. A mutation exchanges glutamate (positively charged, hydrophilic) for valine (uncharged, hydrophobic), which prevents proper association of hemoglobin subunits. Although each subunit can still bind oxygen, they do so poorly; and when deoxygenated, mutated beta subunits aggregate together and cause the erythrocyte to become malformed, or sickled. This leads to the intense pain ("bone crises") patients experience, and can cause vascular damage as well, which leads to autosplenectomy, or spontaneous destruction of the spleen, an organ that normally processes blood cells for removal. The key to this question is that a missense mutation, which exchanges one amino acid for another, can still have highly deleterious effects on protein function. [*Section 5.c*]

8. **(D) Proline hydroxylation**

As will be discussed further in Chapter 9, vitamin C is a key cofactor in the synthesis pathway of collagen. This is a common example used to highlight the importance of post-translational modification: protein function is often incomplete with the primary amino acid sequence alone; rather, further reactions have to take place to assist in proper folding, addition of functional groups, removal of extra sequences, and so on. Collagen requires addition of sugar groups (glycosylation) as well as hydroxyl (OH) groups (hydroxylation). The latter requires vitamin C as a cofactor. In scurvy, which occurs today mostly in the context of severe malnutrition, a diet insufficient in fruits and vegetables, patients present with bleeding of gums and joints, because insufficient collagen synthesis leads to weakened vasculature. [*Section 7.b*]

9. **(A) Aneurysm**

Though it is not stated, this patient likely has Ehlers-Danlos syndrome, which is a disorder caused by mutations in processing collagen into the properly folded triple helix structure. Different types of Ehlers-Danlos are named based on the specific mutation and how it affects collagen throughout the body. The two most common types affect collagen types III and V. Type V is the "classic" subtype (type V collagen associates with type I collagen) and causes joint and skin stretchiness, including hyperextensibility. Type III collagen is involved in vasculature, so these patients are at risk of serious conditions including aneurysm and rupture of vessels due to their weakened structure. [*Section 7.d*]

10. **(E) Mistargeting of lysosomal enzymes**

I-cell disease, also known as mucolipidosis II, is a lysosomal storage disease (such as the classic example, Tay-Sachs). Like these other conditions, I-cell disease results from defective lysosome function. Lysosomes are normally sites of breakdown of various molecules. Lysosomal enzymes, like other cellular proteins bound for specific organelles, are synthesized in the endoplasmic reticulum and processed in the Golgi apparatus. Normally, these enzymes are modified by the addition of a mannose-6-phosphate tag, that marks the enzyme for transport to the lysosome. In I-cell disease, the enzyme responsible for this tag is missing, and lysosomes are thus unable to function. I-cell disease is named for "inclusion cells" which are seen on microscopic examination of cells. Lysosomal enzymes without the tag are excreted from the cell, so blood levels of lysosomal enzymes are increased as well. [*Section 6.b*]

Chapter 4

1. **(B) Movement of myosin along actin decreases**

   This question asks about the organization of the sarcomere. Sarcomeres are made of thick and thin filaments, which move relative to one another to contract the sarcomere and thus the muscle fiber. The thick filament is made of myosin bundles, which are motor proteins. The thin filament is actin, forming the track along which myosin walks. Then, the various zones of the sarcomere are named according to their content. The H band represents the region of thick filaments, while the I band represents thin filaments (H is thicker than I). The A band is the overlap between the two (and contains the H band), representing the region of myosin movement along actin. Thus, the length of the A + I bands makes up one sarcomere. If the width of the H band changes less during contraction, this means the A band is smaller, and myosin is moving less along the thin (actin) filament. [*Section 8.b*]

2. **(D) Myosin filaments do not release actin without ATP**

   Rigor mortis, or the stiffening of the muscles after death, begins between 2 and 4 hours after death, and lasts until muscle fibers decompose, often days after death. It occurs because of the way muscles function on a molecular level. Muscle contraction requires ATP, but the step that requires ATP is actually relaxation of the sarcomere (muscle fiber) and release of the actin (thin) filament by the myosin (thick) filament. (Myosin activity relies on hydrolysis of ATP that has already bound.) Thus, as ATP runs out after death, muscles contract but cannot relax. Only after the muscle fibers themselves degrade over days do muscles again relax. [*Section 8.b*]

3. **(B) Fibrillin**

   This patient, a tall, young athlete who died suddenly of aortic dissection upon physical activity, is a classic presentation of Marfan's syndrome. This is an autosomal dominant disorder, but is an example of variable expressivity, meaning the severity or phenotype can be different in different people. The underlying mechanism is a mutation in the gene expression fibrillin. This protein associates with another protein, elastin, and together they form an important component of the extracellular matrix, which provides structure to all structures within the body. People with Marfan's syndrome are at risk of heart problems such as aortic dissection, as seen here, as well as eye problems, such as lens subluxation (ectopia lentis) which leads to worsening vision bilaterally. [*Section 3.f*]

4. **(D) Passive diffusion through channel proteins**

   This question asks about resting membrane potentials across neurons and other cells. In this study, patient neurons have a higher (more positive) resting potential than neurons from control patients. It is likely that some underlying mechanism is causing either more positive ions (cations) to enter patient neurons, or more negative ions (anions) to leave the neurons. In action potential physiology, a number of voltage-gated channels control flow of ions to propagate a potential. At rest, however, diffusion of potassium (K^+)

cations from the cytosol to the extracellular space is facilitated by so-called "leaky" K^+ channels. This occurs even though the neuron is negative relative to the outside because the chemical concentration of K^+ is much higher within the cell, so the concentration gradient drives its exit. The result is that the resting membrane potential is negative and allows for control of how sensitive a neuron is to an action potential. In patient cells, a higher resting potential likely means that the diffusion of ions is diminished, leading to more K^+ inside the cell and thus more cations inside the cell. [*Section 4.e*]

5. **(B) Cytosol**

Hormone signaling pathways are often tested as part of the Endocrine and Reproductive systems as a way of incorporating Biochemistry. There are three main types of hormone signaling cascades: protein hormones, steroid hormones, and thyroid hormone. Protein hormones are bulky and a mix of hydrophobic and hydrophilic regions, so they general signal by binding to receptors on the plasma membrane and inducing intracellular signaling cascades. Steroid hormones are hydrophobic and small, so they are able to pass through the plasma membrane. In general, they bind to receptors in the cytosol, which facilitate their transport into the nucleus to induce gene transcription. Finally, thyroid hormone binds to its receptor in the nucleus. Thus, estrogen, which is a steroid hormone, will bind to its receptor in the cytosol [*Section 7.e*]

6. **(A) Decreased G protein activation, decreased AC activation**

G-protein signaling is a major pathway by which protein hormones (which bind to plasma membrane receptors outside the cell) induce a signaling cascade within the cell. Parathyroid hormone (PTH) binds a receptor outside the cell and induces signaling via a G-protein cascade (specifically, $G\alpha_s$). Upon activation of G-proteins, a number of downstream effects can happen depending on what type of G-protein is activated. The $G\alpha_s$ pathway induces adenylyl cyclase (AC) to form cyclic AMP, which induces further enzyme activation. The $G\alpha_i$ pathway, in contrast, inhibits AC activity. This patient has symptoms of hypocalcemia, and resultant poor bone growth, but PTH is elevated. This is likely pseudohypoparathyroidism, and is caused by a mutation in the G-protein receptor itself, which leads to dysfunction of the G pathway, though PTH levels and structure is normal. The result is decreased $G\alpha_s$ protein activity and decreased AC activity. [*Section 9.f, Table 4-2*]

7. **(B) It catalyzes cleavage of covalent bonds with the aid of a water molecule**

This question asks about a specific clinical scenario involving a patient with metastatic breast cancer that has spread to the bone. However, the question itself can be answered by just reading the final two lines. Hydrolases have predictable function based on the name: they catalyze cleavage of covalent bonds, splitting up a water molecule into an H and an OH group in the process. In general, knowing the activity of specific enzymes is not high yield for Step 1, but it is helpful to be familiar with the general functions of enzyme classes, in part to help make educated guesses on questions if necessary. [*Section 5.d*]

8. **(E) The mutated enzyme has a lower affinity for glucose-6-phosphate**

Michaelis-Menten kinetics is one method of describing and predicting enzyme activity in a closed system. It is used often in research to study the activity and function of enzymes, and can also help to compare enzymes with mutations to their normal activity, to help identify disease processes. This question asks specifically about the implication of an increase in the K_m of a reaction. The K_m, or Michaelis constant, refers to the concentration of substrate (reactant) of an enzymatic reaction necessary to induce half of the maximal reaction velocity. It is a measure of affinity for the enzyme to its substrate: if a great deal of substrate is necessary, then a lot of the substrate exists not bound to enzyme. Thus, if the K_m is higher, more substrate is necessary for the reaction to proceed at half its maximum velocity, meaning that the mutated enzyme has a lower affinity for substrate. [*Section 6.b*]

9. **(D) Troponin is normally involved in muscle contraction but is released by dying cells after myocardial infarction**

Troponin is a component of the muscle contraction complex, which normally prevents myosin binding to actin but releases upon calcium binding. However, it has also been used as a marker for muscle injury, as different types of troponin exist in different types of muscle. One type of troponin, found within cardiac muscle, can be measured to identify potential recent myocardial infarction. [*Section 8.c*]

Chapter 5

1. **(C) Buildup of acetyl-CoA signals arrest of metabolic pathways used to produce energy**

Acetyl-CoA is a key molecule within metabolism. As outlined in this chapter and detailed in upcoming chapters, acetyl-CoA is involved in the production and breakdown of multiple energy products. It is one of the immediate products of glycolysis, from pyruvate, and is used in the TCA cycle for ATP production via oxidative phosphorylation. It is also produced by catabolism of many amino acids as well as of fatty acids and ketone bodies. As is often the case in metabolism, buildup of intermediate molecules often indicates an error in the pathway. A buildup of acetyl-CoA may indicate that this patient cannot utilize acetyl-CoA properly, or that it is being overproduced by dysfunctional regulation of metabolism. Thus, it is not only concerning that acetyl-CoA metabolism is impaired; it is likely that many other metabolic pathways are impaired as a result. [*Section 5.b, Figure 5-1*]

2. **(B) Lactate decreases, pyruvate decreases, acetyl-CoA increases**

These patients have mitochondrial diseases, and the case states that metabolic function is decreased. Thus, the question is asking what key metabolic processes involve the mitochondria. The major metabolic pathways that involved the mitochondria relevant to Step 1 are oxidative phosphorylation and fatty acid metabolism. In this case, the question asks about lactate and pyruvate, which are both involved in balance with the TCA cycle, meaning this question is asking about oxidative phosphorylation. When oxygen and functional mitochondria are present, pyruvate is converted to acetyl-CoA which then goes through the TCA cycle. When oxygen is not present, or mitochondria are absent or dysfunctional, pyruvate is converted to lactate, which allows for replenishment of cellular NAD^+ stores. In other words, when TCA cycle activity increases due to genetic therapy correcting mitochondrial dysfunction, lactate levels will decrease, and acetyl-CoA will decrease. Overall, less glycolysis will need to take place to produce the same levels of ATP, so pyruvate is hypothesized to decrease as well. [*Section 5.b*]

3. **(B) It inhibits anaerobic glycolysis**

Lactate dehydrogenase is the enzyme that converts pyruvate to lactate, which also converts NADH to NAD^+. This is important in anaerobic metabolism because glycolysis requires NAD^+ to function. Thus, a drug that inhibits lactate dehydrogenase would prevent replenishment of NAD^+ and pyruvate and NAD^+ would build up, both of which would inhibit glycolysis. Of note, NADH can be converted back to NAD^+ via oxidative phosphorylation. However, cancer cells typically utilize glycolysis at a higher rate than normal cells, making this pathway a potential target for anticancer therapies. [*Section 6.b*]

4. **(D) Pyruvate**

Red blood cells are unique when thinking about metabolism because they do not have mitochondria, meaning their only method of producing ATP is via anaerobic metabolic pathways. This means most of an RBC's ATP is produced using glycolysis. Because the final product of glycolysis is pyruvate, it is likely that RBCs would be enriched in pyruvate over the other compounds listed (most of which are involved in fatty acid synthesis, which RBCs do not use). [*Section 5.c*]

5. **(E) The reaction catalyzed by PFK-1 proceeds at a slower rate than any other reaction, so it determines the rate of the overall process**

This question specifically asks about the rate-limiting step of glycolysis. The rate-limited step of a reaction is the step that proceeds at the slowest rate of the various reactions. Thus, the overall reaction can take place no faster than that specific reaction. This also makes the rate-limiting reaction an effective regulator of the pathway overall, and a good drug target if manipulation (speeding up or slowing down) of the overall pathway is desired. [*Section 5.d*]

6. **(E) It is the final electron acceptor of the electron transport chain**

The electron transport chain is a process that takes place in mitochondria as part of oxidative phosphorylation. In this process, the energy of covalent bonds is used to create a proton gradient within the mitochondria. Energy of covalent bonds is captured by a sequence of oxidation-reduction reactions. The specific reactions are not important to know for Step 1. But, it is important to remember that oxygen is required for the electron transport chain to function, because it is the final electron acceptor, being converted to water. Then, ATP synthase converts the proton gradient into ATP production using its electrochemical gradient. [*Section 8.b*]

7. **(E) Inhibition of the electron transport chain**

Cyanide is a highly toxic compound that induces rapid onset respiratory arrest by inhibition of the electron transport chain. A classic presentation may be someone working in a chemical plant, who inadvertently is exposed to a chemical during an industrial process. Symptoms are rapid onset headache, nausea, and dizziness, followed by seizures, loss of consciousness, and cardiac/respiratory arrest. Cyanide binds readily to iron in Complex III of the electron transport chain and prevents oxidative phosphorylation. Treatment with nitrites converts the iron in hemoglobin to a form that binds cyanide more readily than Complex III, thus releasing it from mitochondria (but preventing proper hemoglobin function, a serious but somewhat less immediate problem). [*Section 10.d*]

8. **(D) Glucose-6-phosphatase in kidney cells**

Blood levels of glucose are maintained at stable levels in both feeding and fasting periods by multiple mechanisms. Upon eating, insulin release promotes glucose uptake into cells for use as energy and production of fatty acids, glycogen, and other energy stores. During fasting, glucagon helps to induce breakdown of energy stores and release of glucose into the bloodstream for use around the body. The liver and kidney are major sites of gluconeogenesis. Release of glucose out of liver and kidney cells undergoing gluconeogenesis requires the activity of glucose-6-phosphatase. Other cells such as muscle cells can produce glucose-6-phosphate from intracellular glycogen stores, but owing to the lack of glucose-6-phosphatase they cannot export glucose into the bloodstream for use elsewhere. [*Section 9.a*]

9. **(D) Phosphofructokinase-2**

This question asks about regulation of glycolysis. Insulin promotes glucose uptake by cells from the bloodstream, and then processing of that glucose into ATP, as well as glycogen and fatty acids. Although knowing every enzyme involved in glycolysis is not necessary for Step 1, there are a few key regulatory enzymes that are important to know. The rate-limiting enzyme of glycolysis is phosphofructokinase-1 (PFK-1). The activity of this enzyme is regulated by a second enzyme, PFK-2, which is induced by insulin. Thus, insulin increases PFK-2 activity, which induces PFK-1 and glycolysis over all. The other choices listen here are either not regulated by insulin (hexokinase) or are actually involved in gluconeogenesis, the production of glucose from other metabolites. [*Section 5.d*]

10. **(D) GLUT-4**

There are many different glucose facilitators/transporters in the body, and each has a different distribution within tissues to allow for careful control of glucose intake depending on that cell's needs. However, a key point to remember is that only one major subtype is regulated by insulin: GLUT-4. This is found primarily in adipose tissue, to induce fatty acid synthesis during feeding and abundant glucose; and striated muscle, to allow muscle cells, which have high-energy requirements, control over energy depending on availability. GLUT-2 is another important glucose facilitator as it is found in the pancreas, the site of insulin production and one of the first parts of the body to respond to high glucose intake during feeding. [*Section 4.a, Table 5-1*]

Chapter 6

1. **(D) Hereditary fructose intolerance**

   This patient experiences symptoms when eating fruit-based foods, so the disorder is likely in fructose metabolism, which as its name suggests, is found in fruits and other naturally sweet foods such as honey. There are two metabolic disorders of fructose intolerance to know for Step 1: essential fructosuria, which has no major symptoms except fructose in the urine; and fructose intolerance, which can lead to symptoms such as vomiting and hypoglycemia, as in this patient. [*Section 4.f*]

2. **(E) Galactose 1-phosphate uridyltransferase**

   This infant experiences GI symptoms after breastfeeding, indicating a mutation in galactose metabolism. Lactose, primarily found in milk, is a disaccharide made up of glucose and galactose, making this the common source of symptoms in patients with disorders of galactose metabolism. Like fructose metabolism, there are two main conditions to know. The first, galactokinase deficiency, primary presents as cataracts, but not GI symptoms. The more serious one, classic galactosemia, has a mutation in the enzyme uridyltransferase, leading to cataracts, but also GI and liver symptoms. [*Section 5.f*]

3. **(D) Sorbitol**

   Sorbitol is a product of alternate glucose metabolism that, like processing with hexokinase, serves to trap glucose in the cell. However, some tissues, such as the retinal and lens, have limited ability to process sorbitol. When it builds up, it can cause osmotic swelling and cell damage. Clinically, this is typically seen in diabetics, who often have high circulating blood glucose levels. Although advanced glycation end-products also cause damage such as neuropathy and arthritis, the cataracts in this patient are most likely linked directly to sorbitol, which is a separate biochemical process. [*Section 6.d*]

4. **(B) Debranching enzyme**

   This question asks very specifically about the biochemistry of glycogen. Glycogen synthesis involves long chains of glucose molecules joined by α-(1,4) glycosidic bonds (meaning carbon atoms 1 and 4 are linked). To increase density, some glucose molecules also have α-(1,6) glycosidic bonds, which allows a "branch" of additional glucose molecules to build of the core chain. In glycogen breakdown, debranching enzyme cleaves these α-(1,6) glycosidic bonds, so increased numbers of α-(1,6) bonds relative to α-(1,4) indicates inhibition of debranching enzyme. Choices A and E are involved in glycogen synthesis, while choices C and D are involved in other steps of glycogen breakdown. [*Section 7.d*]

5. **(C) McArdle disease**

   This patient complains of muscle cramps that occur with intense activity, and myoglobinuria is noted. These are the classic presentations of McArdle disease, which is a specific deficiency of glycogen breakdown within muscle cells. The disorder is in a glycogen phosphorylase specific to skeletal muscle, so McArdle disease lacks the other findings (such as hepatomegaly) common to many other glycogen storage diseases like Von Gierke and Cori disease. [*Section 7.e, Table 6-2*]

6. **(C) Pompe disease**

   Muscle weakness in a young infant may seem at first like muscular dystrophy, a genetic disorder often tested on Step 1. However, this question specifically identifies lysosomal glucose production as the source of the defect. Because the lysosome is specifically targeted, this is Pompe disease, which is caused by a specific defect in lysosomal breakdown of glycogen. (Knowing the exact enzyme is not high yield, but it is a special glucosidase that can break down both α-(1,4) and α-(1,6) bonds, also called acid maltase.) This has a characteristic presentation among glycogen storage diseases because it affects the heart, leading to cardiomegaly, in addition to systemic effects including liver damage. [*Section 7.e, Table 6-2*]

7. **(E) Von Gierke disease**

   This young patient presents with a classic presentation of Von Gierke disease, the most common glycogen storage disease. She presents with weakness, hypoglycemia, hepatomegaly, and lactic acidosis. However, it would be difficult to confidently identify Von Gierke versus other diseases such as Cori disease (even knowing that Von Gierke tends to be more serious) without the additional information that glucose-6-phosphatase is deficient in this patient. Mutations in this enzyme, the final step in glycogen breakdown that releases free glucose, cause Von Gierke disease. [*Section 7.e, Table 6-2*]

8. **(D) Sulfamethoxazole**

   This patient presents with methicillin-resistant *Staphylococcus aureus*, or MRSA. This drug is resistant to many common antibiotics, but the question specifically asks which drugs should be avoided based on the patient's diagnosis of glucose-6-phosphate dehydrogenase (G6PD) deficiency. Patients lack a functional pentose phosphate pathway (HMP shunt), meaning they have reduced ability to produce NAPDH, a free radical scavenger. Sulfa drugs should be avoided in patients with G6PD deficiency, as they can increase the risk of hemolytic anemia. [*Section 8.e*]

9. **(E) Scavenge free radicals to prevent reactive oxygen species formation**

   Free radicals are harmful molecules that have unpaired electrons, making them highly reactive and thus potentially damaging to cells, both animal and plant-based. In biology, free radicals are often oxygen-based. The body uses free radicals to its benefit in some cases, such as in the respiratory burst, and also has mechanisms to protect against free radicals. The general term "antioxidant" refers, as the term suggests, to compounds or chemicals that neutralize reactive oxygen species, in the same way the body's own processes do. This prevents damage to cells and, as in this question, can help preserve food from spoiling. This explains why citrus juice, which contains the antioxidant vitamin C, prevents browning (oxidation) of sliced apples. [*Section 8.c*]

10. **(E) Increased erythrocyte sensitivity to free radicals disrupts Plasmodium life cycle**

    The pentose phosphate pathway (HMP shunt) is one of the body's main mechanisms to produce NADPH, which is both a source and scavenger of free radicals in the body. The pentose phosphate pathway is active in red blood cells, which are also the targets of the parasitic *Plasmodium* protozoa, the causative organisms of malaria. People with glucose-6-phosphate dehydrogenase (G6PD) deficiency display resistance to malaria because red blood cells are highly sensitive to free radicals and thus undergo hemolysis more readily. The reduced life span of red blood cells makes it more difficult for malarial organisms to complete their life cycle and proliferate. [*Section 8.e*]

Chapter 7

1. **(B) Glutamate → Valine**

   Amino acids are characterized by the properties of their side chain groups (R groups). First, they may be polar or nonpolar, with nonpolar amino acids more likely to associate with hydrophobic environments (such as in a transmembrane domain spanning a phospholipid bilayer). Second, polar amino acids may be charged or uncharged. Mutations that change a charged, polar amino acid into a nonpolar amino acid dramatically shift the properties of that amino acid, which has the highest likelihood to alter the folding and function of the protein. Thus, glutamate (charged) to valine (nonpolar) is the most significant mutation here, and indeed, is a mutation seen in patients with sickle cell anemia. [*Section 2.c*]

2. **(E) Lysine, phenylalanine**

   Among the naturally occurring amino acids, some can be synthesized by processes within the human body, while others can only be obtained from the diet. Though learning all properties of amino acids is not high yield for Step 1, knowing which amino acids are essential is useful, along with which amino acids are polar/nonpolar and charged. In this case, only lysine and phenylalanine are both essential. [*Section 2.e*]

3. **(E) Pyruvate dehydrogenase**

   This patient likely has pyruvate dehydrogenase deficiency, which causes lactic acidosis and neurologic symptoms. Of the choices listed, two are potential answers. Pyruvate dehydrogenase links glycolysis to the TCA cycle by converting pyruvate into acetyl-CoA. Phosphofructokinase-1 is the rate-limiting step of glycolysis; however, it is so central that mutations are very rare. The other choices are involved in glycogen breakdown (A), fatty acids (B), and purine nucleotides (C). [*Section 3.c*]

4. **(E) Tyrosine hydroxylase**

   Parkinson's disease, as noted, results from decreased dopamine production in the substantia nigra. Dopamine is one of the catecholamine neurotransmitters (along with norepinephrine and epinephrine), which are produced from amino acid precursors. The first step of the pathway is the conversion of phenylalanine to tyrosine (the step mutated in phenylketonuria), which is then converted into the precursor L-dopa by tyrosine hydroxylase. These two steps are key steps to know for Step 1. The other answer choices list enzymes involved in catecholamine breakdown (B and C) or in other amino acid metabolism pathways (A, associated with maple syrup urine disease; and D, associated with homocystinuria). [*Section 4.c*]

5. **(C) Glutamate ⟶ GABA**

   Vitamins will be covered in additional detail in Chapter 9. However, some vitamins play important roles in neurotransmitter synthesis, and deficiencies can lead to neurologic symptoms. Of these, vitamin B_6 is especially important, as it is a cofactor for synthesis of dopamine, histamine, GABA, and serotonin. It is not directly involved in norepinephrine synthesis, which is synthesized from dopamine. [*Section 4.d, Figure 7-2*]

6. **(B) High serum orotic acid**

   This patient has a urea cycle disorder. The most common disorder associated with the urea cycle is ornithine transcarbamylase (OTC) deficiency, an X-linked disorder that causes hyperammonemia and associated neurologic and liver damage. Mutation of OTC leads to buildup of orotic acid, which is exacerbated after eating a meal containing protein. The other choices either list compounds downstream of the deficiency (A, C) or those unrelated to the urea cycle (D, E). [*Section 5.d*]

7. **(D) Tyrosine**

   "Mousy odor" in a child is a buzzword for phenylketonuria (PKU). It is particularly common in families that have immigrated to the United States, where it is routinely tested for at birth. PKU results most commonly from a deficiency of phenylalanine hydroxylase, which prevents conversion to tyrosine. Treatment involves not only restriction of phenylalanine from diet (though not completely eliminated as it is an essential amino acid) but also supplementation with tyrosine, which under conditions of limited phenylalanine availability becomes essential and must be obtained from diet. [*Section 6.a*]

8. **(D) Urine that turns black in air**

   Urine that turns black in air is associate with alkaptonuria, a benign condition that results from homogentisate oxidase. A mousy smell (choice A) is associated with phenylketonuria, elevated urine homocysteine levels (choice B) indicate homocystinuria, elevated orotic acid levels (choice C) indicate OTC deficiency (urea cycle disorder), and urine with a maple syrup odor (choice E) indicate maple syrup urine disease. All of these other choices are more serious than alkaptonuria. [*Section 6.b*]

9. **(B) Isoleucine**

Just as a "mousy odor" is characteristic of PKU, urine with a sweet or syrupy smell is indicative of maple syrup urine disease, which is, in short, a deficiency in metabolism of branched chain amino acids. The branched chain amino acids are isoleucine, leucine, and valine (all nonpolar). Thus, only isoleucine, among the choices given, would be restricted from this infant's diet. [*Section 6.c*]

10. **(C)**

Kidney stones in children are relatively rare, and cystinuria is the most common cause. Like other types of kidney stones, the presentation is often unilateral, sharp flank pain (lower side/back). Crystals form because excretion of the amino acids is inhibited. Thus, concentration builds up and crystals precipitate in the acidic environment of the urine. Thus, treatment involves reducing concentration of dibasic amino acids, and includes increasing fluid intake, decreasing protein and sodium, and increasing pH. [*Section 6.e*]

Chapter 8

1. **(A) Acetyl-CoA → Malonyl-CoA**

This question is asking which of the reactions are part of a process induced by insulin. In addition to inducing cellular glucose uptake, insulin generally promotes anabolic processes, including glycogen synthesis and fatty acid synthesis. Of the reactions listed, only choice A is part of one of these process (as the first step of fatty acid synthesis). Choices B and E occur in gluconeogenesis, C in glycogenolysis, and D in fatty acid breakdown. [*Section 3.b, Figure 8-2*]

2. **(D) Inability to break down long-chain fatty acids**

The physician believes this patient has systemic primary carnitine deficiency, which is caused by a defective carnitine shuttle system from cytosol into mitochondria. The result is that long-chain fatty acid metabolism is impaired, since they cannot be imported into the mitochondria for breakdown into acetyl-CoA. Patients can experience weakness and hypotonia that resemble other muscle weakness disorders, but it is primary due to lack of energy; hepatomegaly and hypoketotic hypoglycemia are also seen. Only the correct answer is involved in fatty acid metabolism; others refer to different processes that involve mitochondria. [*Section 3.e*]

3. **(C) Inhibition of HMG-CoA reductase**

Only one drug, simvastatin, is specifically given to reduce cholesterol levels. Statins are HMG-CoA reductase inhibitors, which inhibit cholesterol synthesis by the liver, which in turn must uptake more cholesterol from the circulation, reducing LDL levels. Choice A is an enzyme involved in cholesterol breakdown, choice B describes the mechanism of fibrates such as gemfibrozil, choice D is the mechanism of niacin, and choice E would have the opposite effect, preventing LDL removal from the bloodstream. [*Section 4.d*]

4. **(D) ApoE4**

   Most cases of Alzheimer's disease have not been linked to specific inherited genes, though population data suggests that there is a genetic component. However, some genetic variants have been described that are linked to increased risk of the disease, and one of the most well-known is that of the ApoE variant ε4. There are three variants of ApoE, and the number of copies (up to two) a person has predicts increased risk of developing Alzheimer' disease. In contrast, the variant ApoE ε2 has the lowest risk of Alzheimer's with ApoE ε3 in the middle. [*Section 5.h*]

5. **(C) VLDL**

   This patient is taking a statin, which inhibits HMG-CoA reductase. In normal lipid transport, there are two base lipid-carrying particles: chylomicrons are absorbed from the GI tract, which are reduced to chylomicron remnants as lipids are absorbed by peripheral tissues; and VLDL, which is released by the liver and converted eventually into IDL and then LDL after uptake of lipids by peripheral tissues. [*Section 5.c*]

6. **(B) Apo B-48**

   This patient may have abetalipoproteinemia or a related disease. Apo B-48 is important for chylomicron uptake from the GI tract, and without it, fatty acids and fat-soluble vitamins cannot be absorbed, leading to severe symptoms beginning early in life. Steatorrhea (fat in stool) and severe failure to thrive are symptoms. [*Section 5.g, Table 8-2*]

7. **(E) Lipoprotein lipase**

   Lipid transport is complicated, but some enzymes in particular are important to know for Step 1. The most important is HMG-CoA reductase, the site of action of statins, one of the most widely used drug classes. Lipoprotein lipase (LPL) is the target of fibrates, another lipid-lowering drug class. LPL acts to remove triglycerides from circulating chylomicrons and VLDL particles, thus converting them into chylomicron remnants and IDL particles. Apo B-100 and ApoE both bind LDL receptor. [*Section 5.d*]

8. **(E) Removes cholesterol from tissues**

   High density lipoprotein (HDL) is released from the liver and is not a product of LDL or chylomicron metabolism. It removes cholesterol from peripheral tissues and recycles it to the liver. Also called "good cholesterol" in popular vernacular, it is associated with a decreased risk of coronary artery disease. The other choices listed are mechanisms that would lower peripheral cholesterol levels and thus reduce atherosclerosis, but they are not the mechanisms of HDL. The drug described in the case is likely in the fibrates class, along with gemfibrozil, fenofibrate, and others. [*Section 5.f*]

9. **(C) ApoC-II**

   This patient presents with signs of hepatosplenomegaly and acute pancreatitis (as noted by the elevated serum amylase levels). Furthermore, the yellow spots on the patient's back (known as xanthomas) are indicative of high lipid levels. Together, these indicate a possible familial hyperlipidemia. In particular, these symptoms are consistent with dyslipidemia type I, also known as hyperchylomicronemia. In this condition, activity of lipoprotein lipase (LPL) is decreased, either due to a mutation in LPL directly or a mutation of ApoC-II, its cofactor. The result is an inability to process chylomicrons in the circulation, and thus buildup of lipids within the blood stream, which accumulate in peripheral tissues and cause the symptoms seen here. [*Section 5.g*]

10. **(C) Gemfibrozil**

This patient has tried various lipid-lower agents with minimal success. Additional options are available for such patients, though they often have more side effects. The endocrinologist here recommends a drug that activates the enzyme cholesterol 7α-hydroxylase, which in turn increases HDL synthesis. (It also upregulates LPL activity to reduce LDL and triglyceride levels.) This describes the class of fibrate drugs, including gemfibrozil. [*Section 5.g*]

Chapter 9

1. **(D) Metabolism of fatty acids leading to acidosis**

This patient is likely experiencing diabetic ketoacidosis, which is the result of poorly controlled diabetes. More common in type I diabetics, particularly those with undiagnosed type I, patients do not produce enough insulin. This leads to fatty acid metabolism, causing ketosis, while also leading to diuresis as glucose spills into the urine. [*Section 2.d*]

2. **(E) Pyruvate**

This case is describing normal lactic acidosis that results from strenuous exercise and resulting anaerobic metabolism. In the absence of oxygen, pyruvate is metabolized to lactic acid to replenish NAD^+ stores. Lactic acid buildup leads to the burning sensation in muscles during prolonged exercise. In more sustained but less intense exercise, blood flow is able to wash out lactic acid and prevent buildup, and oxygen can reach tissues to allow for some aerobic metabolism. [*Section 3.d*]

3. **(D) Muscle wasting**

Severe malnutrition can be divided into two categories based on the constellation of symptoms. In kwashiorkor, substantial peripheral edema and central swelling are present, along with general wasting. In marasmus, there is not substantial edema, but rather extreme muscle wasting, and patients are very thin over all. There are hypothesized pathologic mechanisms behind these differences, but it is likely there is overlap between the two. Here, though, the presence of muscle wasting is the only one of the symptoms listed that is consistent with marasmus. The first three choices are more consistent with kwashiorkor, while poor night vision is characteristic of vitamin A deficiency. [*Section 4.d*]

4. **(A) Vitamin A**

Patients with unstable living conditions and access to food are at increased risk for nutritional deficiencies. This question asks about persistent skin rash and infection. Skin rashes are consistent with a number of vitamin deficiencies. Vitamin A is typically associated with night blindness but can also lead to scaly skin. It also causes immune deficiency, leading to persistent infections. Thus, it is the most consistent answer. Vitamin B$_3$ can also present as a photosensitive skin rash, but the classic triad of dementia and diarrhea along with dermatitis are not seen here. The other choices are not associated with skin rashes. [*Section 6.b*]

5. **(C) Osteomalacia**

This patient's symptoms are classic presentations of rickets, a disease in children caused by vitamin D deficiency which leads to improper bone growth prior to the closure of epiphyseal plates. Vitamin D plays a role in calcium and phosphate regulation, which also underlies the patient's other symptoms. In adults, vitamin D deficiency does not cause misshapen bones as they have finished growing; but it does lead to weakening of bones in a condition called osteomalacia, which involves bone pain and increased risk of bone fractures, as well as compensatory (secondary) hyperparathyroidism. [*Section 6.c*]

6. **(E) Vitamin K**

Difficulty clotting in a newborn is the characteristic presentation of vitamin K deficiency. Vitamin K is synthesized with the help of intestinal bacteria, which newborns do not yet have. Thus, newborns are given vitamin K injections to provide this until they can synthesize their own. Vitamin K is important for several clotting factors, and thus deficiency leads to increased bleeding and bruising. [*Section 6.e*]

7. **(C) Vitamin B$_1$ (thiamine)**

A chronic alcohol user with neurologic symptoms such as ataxia and confusion is the classic presentation of Wernicke-Korsakoff syndrome, or chronic vitamin B$_1$ (thiamine) deficiency. Thiamine is involved in ATP production pathways, particularly those involving glucose metabolism (glycolysis, TCA cycle). Deficiency leads to two major conditions: beriberi in malnutrition, and Wernicke-Korsakoff syndrome in alcoholism. Alcohol users often are malnourished as alcohol does not contain key nutrients. In addition to this patient's confusion and ataxia, this syndrome can include short-term memory loss and ophthalmic symptoms. [*Section 7.d*]

8. **(B) Hydroxylation of proline**

This patient's significant bruising and gingivitis, coupled with a history consistent with malnutrition, suggest vitamin C deficiency and a diagnosis of scurvy. Vitamin C is important for hydroxylation of proline and lysine during collagen synthesis. Because collagen is an important support structure within vasculature, deficiency in collagen leads to vascular weakness and resulting bleeding. That the patient is homeless and has an unknown medical history suggests vitamin deficiency is possible; in a patient that can provide history of a balanced diet, additional factors may be considered, such as a genetic clotting disorder. [*Section 7.l*]

9. **(B) Neural tube defect**

Neural tube defects, such as spina bifida, are disorders of early fetal development that result from a failed closure of the neuropore, exposing the spinal column to the amniotic fluid. Folate deficiency increases the risk of neural tube defects, and thus folate is recommended as supplementation during pregnancy. It is found naturally in green, leafy foods such as spinach. Patients that have not accessed healthcare for their pregnancy should be specifically asked about vitamin supplementation, which may have been the case for this patient's previous pregnancy. [*Section 7.j*]

10. **(E) Megaloblastic anemia**

Vitamin B_{12} deficiency is rare, as it can be stored in the liver for years. However, it can be seen in patients with autoimmune disorders (pernicious anemia, Crohn's disease) and those with prolonged nutritional deficiencies. Patients with vitamin B_{12} deficiency present with megaloblastic anemia due to its role in DNA synthesis, as well as paresthesias and gait disturbances, as it is important in myelination of the nervous system. Iron deficiency can also be a cause of anemia, but it is through a different mechanism unrelated to vitamin B_{12}; and ataxia is more characteristic of vitamin B_1 deficiency as part of Wernicke-Korsakoff syndrome. [*Section 7.k*]

Chapter 10

1. **(B) $G_1 \rightarrow$ S cell cycle transition**

This patient has retinoblastoma, a cancer of retinal cells that is almost always seen in young children. When shining a light on the eye, it appears as an abnormal reflection. Bilateral abnormalities like this patient's are almost always due to genetic mutation of the *RB1 (Rb)* gene, which produces a protein that serves as a checkpoint in the cell cycle. This is an example of a tumor suppressor gene, as the child likely was born with one good copy of the gene, but loss of the only functional copy in a few cells led to early development of retinoblastoma. [*Section 5.d*]

2. **(A) IGF-1 receptor activating mutation promoting cell division**

Oncogenes and tumor suppressor genes are two classes of genes in which mutations promote the development of cancer. Tumor suppressor genes describe genes whose products generally prevent disordered cell growth, with functions in apoptosis, regulation of the cell cycle, and so on. Oncogenes, in contrast, generally promote cell growth in their normal function, such as growth factor receptors. Mutation of one copy of an oncogene can promote cancer due to unregulated pro-growth signals. Thus, of the choices listed, an IGF-1 receptor activating mutation that promotes cell division describes an oncogene mutation. [*Section 6.d*]

3. **(E) Receptor tyrosine kinase signaling**

Herceptin, or trastuzumab, is an antibody treatment designed against the HER2/neu receptor tyrosine kinase growth factor receptor, which is encoded by a proto-oncogene. Mutations in this receptor contribute to development of aggressive breast cancer (and gastric cancer). Thus, when examining the molecular profile of a cancer, if activating mutations in the receptor are found, Herceptin is a potential treatment option that has seen a high degree of success. [*Section 6.f*]

4. **(D) 47,XX(+21)**

   The results given here are characteristic of Down syndrome, the only trisomy in which β-hCG (and inhibin A) is increased. The exact mechanism of this unique increase is not fully understood, but it is a useful, noninvasive early screening tool to identify potential trisomies early in the pregnancy. [*Section 8.g*]

5. **(A) Autosomal dominant**

   The clinical background of this patient is not necessary to answer this question, as the pedigree provides all of the necessary information. That an affected father passes the condition to a daughter and son eliminates the possibility of X-linked disease. The disease is present in about half of the children of affected parents in each generation and thus is likely autosomal dominant. [*Section 10.b*]

6. **(E) X-linked recessive**

   The question stem states that the dementia results from a single-gene mutation which would be necessary to use the pedigree to assign a mode of inheritance to this family's condition. Otherwise, the pedigree is sufficient to answer the question without background knowledge of the disease. Note that the disease skips generations, and can show up in a child but neither of their parents. This indicates recessive inheritance, and because the father passes it to both sons and daughters, it cannot be X-linked recessive. It is not mitochondrial because it is passed down from the father, while mitochondrial disease is only passed on from the mother. [*Section 10.e*]

7. **(C) 4%**

   This requires the use of the Hardy-Weinberg equation. The prevalence, 1/2500, is equal to q^2. Thus, $q = \sqrt{1/2500} = 0.02$ and $2pq = 2(0.02)(\sim 1) = 4\%$. Note that we can approximate p, the allelic frequency of the normal allele, as 1, because the answer choices vary by factors of 2, much larger than the difference between 0.98 and 1. [*Section 14.c*]

8. **(B) 1.5%**

   This question requires calculating the chance that the red child's father is a carrier. We know that the father's parents are both carriers as they have an affected daughter, making the Punnett square as follows:

AA	Aa
Aa	aa

   However, the father cannot be affected, so his chances of being "aa" are 0. This means he has a 2/3 chance of being a carrier. Then, we must calculate the chance that the mother is a carrier based on the prevalence of the disease of 1/400. Using the Hardy-Weinberg equation, $q = \sqrt{1/400} = 0.05$ and $2pq = 2(0.05)(\sim 1) = 0.1$. Finally, the chance that they pass on the disease to their child (based on the Punnett square above) is 0.25. Multiplying these probabilities together gives the answer, because all three events must happen at the same time: $(2/3)(0.1)(0.25) = 1.67\%$. [*Section 14.c*]

9. **(D) Hexosaminidase A**

This question can be answered from the final sentence alone. It is testing knowledge of the biochemistry behind Tay-Sachs disease, the most common lysosomal storage disease. Tay-Sachs, which causes neurodegeneration and early death, is the result of a mutation in hexosaminidase A. This leads to buildup of GM2 ganglioside, a fatty acid compound, in neurons and progressive neuronal loss. Of the other choices, choice A is mutated in Fabry disease, choice C in Gaucher disease, and choice E in Niemann-Pick disease. (Choice B is not associated with a lysosomal storage disease.) [*Section 13.e, Table 10-5*]

10. **(A) ~0%**

This is perhaps a trick question, as no calculation is necessary. Fabry disease is one of the few lysosomal storage diseases (along with Hunter syndrome) that has X-linked recessive inheritance. Thus, any male born with an affected X chromosome will have the disease. Because the husband in this couple does not have Fabry disease, it does not matter that he has a sister who does; he cannot pass it on. The wife is in a similar situation, as her father was presumably not affected, so it does not matter that it was present in his extended family. Thus, the chance that their child is affected is near zero. [*Section 13.e, Table 10-5*]

11. **(B) Hepatosplenomegaly**

The cherry-red macula is common to many lysosomal storage diseases, the result of buildup of lipids that cannot be metabolized in retinal ganglion cells. However, other symptoms differ slightly between diseases. Tay-Sachs, the most common, does not have hepatosplenomegaly, but Niemann-Pick disease does. This is the main feature that distinguishes the two; Niemann-Pick disease also has foam cells (macrophages) visible on histology. [*Section 13.e, Table 10-5*]

Index

Page numbers followed by f and t indicate figures and tables respectively.

Access videos that provide additional explanations of key high-yield content. All content is high yield in order to pass the USMLE® Step 1 Exam. Videos on key high-yield topics break down and explain concepts in 5- to 7-minute segments.

mhprofessional.com/usmle-biochemistry-review

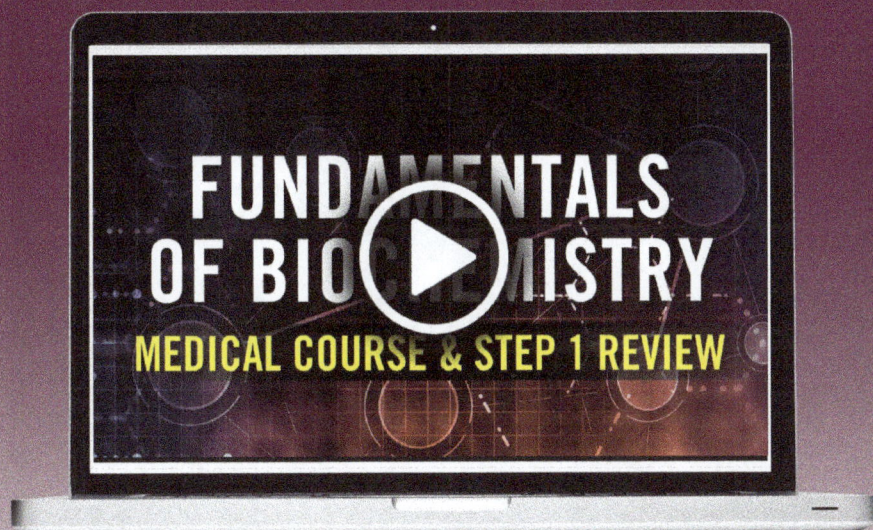

David DiTullio, author of *Fundamentals of Biochemistry: Medical Course & Step 1 Review,* is a recent medical graduate from the UCLA David Geffen School of Medicine and also spearheaded UCLA's Peer Tutoring Program, which offers students who need extra help the chance to partner with a mentor. David also was responsible for providing weekly study sessions for the USMLE Step 1 to review material and provide tips on how to stay organized.

Esteben C. Dell'Angelica, PhD, is the Course Director for Biochemistry, and Professor of Human Genetics at UCLA.

McGraw Hill Education

www.ingramcontent.com/pod-product-compliance
Lightning Source LLC
Chambersburg PA
CBHW081533220326
41598CB00036B/6419